Venus and Mars

The World
of the Medieval Housebook

Christoph Graf zu Waldburg Wolfegg

Venus and Mars

*The World
of the Medieval Housebook*

Prestel   Munich · New York

This book has been published in conjunction with the exhibition held at the
National Gallery of Art, Washington D.C. (November 8, 1998 – January 31, 1999)
and The Frick Collection, New York (May 9 – July 25, 1999).

© Prestel-Verlag, Munich · New York, 1998

Photographic Credits: see page 115

Front cover: Detail from *The Bathhouse* (fol. 18v)
Back cover: Details from *The Children of the Planets: Mars* (fol. 13r), and *Venus* (fol. 15r)
Frontispiece: Detail from *The Mining Panorama* (fol. 35r)

Translated from the German and edited by Almuth Seebohm, Munich

Prestel-Verlag
Mandlstrasse 26 · D-80802 Munich, Germany
Tel. (89) 38 17 09-0; Fax (89) 38 17 09-35
and 16 West 22nd Street, New York, NY 10010, USA
Tel. (212) 627-8199; Fax (212) 627-9866

Prestel books are available worldwide.
Please contact your nearest bookseller or write to either of the above addresses
for information concerning your local distributor.

Lithography by Karl Dörfel, Munich
Typeset by Reinhard Amann, Aichstetten
Typeface: Bembo
Printed by Buchdruckerei Holzer, Weiler im Allgäu
Bound by Oldenbourg, Kirchheim bei München

Printed in Germany on acid-free paper

ISBN 3-7913-1839-X

Preface

The Medieval Housebook has been a carefully safeguarded treasure at Wolfegg Castle ever since the 17th century. Such precious items have laws of their own: on the one hand there is the frequently expressed desire to view the original, on the other the concern to preserve and keep it for future generations.

It is now exactly forty years ago that my uncle, Dr. Johannes Graf zu Waldburg Wolfegg, undertook to write a monograph on the Medieval Housebook and convey its fascination to an interested audience. Today Christoph Graf zu Waldburg Wolfegg is following in his footsteps, while taking advantage of new reproduction techniques. Dismantling the codex for the recently published facsimile provides the opportunity to exhibit all of the pages in a few selected museums. The present volume offers a lively and vivid introduction to the world of the Medieval Housebook for the occasion. "Venus and Mars" stand not only for the planetary deities in the Housebook but also for the antitheses in it, as it includes both the varieties of courtly love and the technology of warfare.

My thanks go to everyone involved, especially Margret Stuffmann, Jan Piet Filedt Kok, Christoph Vitali, Timothy Husband, Andrew Robison and Samuel Sachs. I hope that the circle of devotees fascinated by the Medieval Housebook will grow even larger.

Johannes Erbgraf zu Waldburg Wolfegg

The Medieval Housebook

Only very few works of art succeed in giving the impression that something which took place 500 years ago just happened yesterday.[1] One of these is the so-called Medieval Housebook of Wolfegg Castle with its incomparably vivid representations of the world of the late 15th century.

The appellation of 'Housebook' was given to the manuscript because it was held to be a kind of manual covering all matters of importance for the household of a knight's castle. The subjects covered, however, extend far beyond everyday household uses: they include not only remedies against constipation and recipes for making soaps but also instruction in warfare and defense, as well as sections on jousting and courtship, and treatises on the art of memory and astrology. The book thus encompasses broad spheres of knightly life and of the late medieval intellectual world.

It was in the 19th century that the manuscript acquired the name by which it is commonly known: in 1865 Ralf von Retberg called it "a 'medieval housebook' because it contains miscellaneous matters which might seem of importance to the owner of a household, or, more specifically, a castle, as well as other matter he [...] deemed worthy of note."[1]

This term was then adopted for the unknown artist, who was given the provisional name of 'Housebook Master,' or 'Master of the Amsterdam Cabinet' after the repository of the majority of his engravings. The Housebook Master was presumably active in the Middle Rhine area. Only one documented date appears in his oeuvre: 1480 is on a title page he drew for a romance for the Heidelberg court.[2] He never signed his work, not even with a monogram, which would have been common practice at the time. A number of paintings are attributed to him or his circle – best known are the panel of a *Pair of Lovers* in Gotha and the so-called Speyer Passion Altarpiece – but his work in the graphic arts is his claim to fame. His preferred medium was drypoint engraving, which was not used by his contemporaries. This involves scratching directly into the metal printing plate with the drawing needle, a technique which granted the greatest possible artistic freedom. Its drawback, however, was that only a limited number of impressions could be made because the lines on the plate wore off quickly.

Questions of attribution remain controversial. Whether and to what extent the works subsumed under the Housebook Master's name are by a single master, or whether several artists must share credit for

1 *The coat of arms of the patron, fol. 2r*

the works known so far, has long been a subject of debate among art historians. According to some specialists, the Amsterdam engravings and only a few of the drawings in the Medieval Housebook are still considered to be masterpieces by the artist himself. The rest of the Housebook's drawings are said to be by various other artists.[3] In opposition, I would argue for the attribution of most of the drawings in the Housebook to a single artist. I shall explain why below.

The oeuvre of the Housebook Master is unusual in that it encompasses a wide range of primarily secular themes, from lovers to fanciful coats of arms. Unlike many of his contemporaries, the artist knew how to lend his figures vitality. Whether in childish play or deadly earnest, they are always very moving in their immediacy.

2 The children of the sun, detail, fol. 14r

3 *The collector Maximilian Willibald,*
copperplate engraving by Peter Clouwet
after a portrait by Anselm van Hulle, 1655

The Housebook is preserved at Wolfegg Castle, which is in the Allgäu region in the land of Baden-Württemberg in southern Germany. This is the ancestral castle of the older branch of the Waldburg family (the second branch being the Waldburg-Zeil family), whose principality lasted up to the end of the German Empire. The Waldburg coat of arms shows three lions. They were taken over from the Hohenstaufen emperors, under whom the Waldburgs occupied a number of distinguished leading positions, including that of lord high steward. Being one of the appointments at court under the empire, it later became hereditary, which led to the monstrosity of a title of Imperial Hereditary Lord High Steward. The basic function on which the status of this appointment was based consisted of carrying the imperial orb at the coronation of the emperor. Hence the orb is depicted on the Waldburg coat of arms. One of the hereditary lord high stewards, Maximilian Willibald (1604-1667), laid the foundations for the collection of prints and drawings at Wolfegg Castle (fig. 3).[4] He was an imperial officer, commander of Bregenz and later of the fortress of Lindau. Keeping the troops equipped during the Thirty Years' War was a nerve-wracking task, which he carried out with increasing reluctance. As an imperial officer, Maximilian Willibald successfully defended Lindau against the Swedes, who, however, set Wolfegg Castle on fire. Through his good connections as a privy councillor and chamberlain of the Bavarian Prince Elector Maximilian, he came to be the governor of Amberg for the Upper Palatinate, which had fallen into the hands of the Munich line of the Wittelsbachs. He stayed there for the rest of his life. So far, his career sounds rather provincial and limited to southern Germany. Quite the contrary was the case. He often travelled on diplomatic missions. Besides the obligatory French, he spoke fluent Italian and after the death of his first wife (née Hohenlohe-Waldenburg, in 1645) he married Countess Clara Isabella von Arenberg in Brussels. Fully in keeping with the tastes of the age, he was interested in natural science and maintained relations with the chemist and alchemist Johann Rudolf Glauber (1604-1670). His real enthusiasm, however, was for collecting works of art. He commissioned numerous paintings and, as he had little opportunity to escape from his duties, he appointed agents to be on the lookout everywhere for suitable items.

In fact it was not his military and diplomatic achievements that were to live on but his passion: his collection. He had acquired a French prayerbook with many prints as early as his student days, but it was only from 1644 on that he began to satisfy his collecting desires to the full. Even while seeking a bride he asked his matchmaker to keep an eye out for works of art for his collection. Contemplating a precious biblical manuscript, "his teeth became long and his appetite grew" and he considered selling a good horse for it. He obviously spent an enormous sum on his passion. Apart from that, he exhibited a thriftiness bordering on miserliness. But he had to, seeing as his financial situation was so desperate. The Thirty Years' War and its consequences, the reconstruction of Wolfegg, and his external post in Amberg were heavy burdens. Maximilian Willibald faced them with, among other things, a plea for deferment of payment from his debtors.

In his will he bequeathed his collection as a fideicommission, inalienable family property, and stipulated that a catalogue be made. The inventory was not completed until five years after his death, in 1672. Everything was to remain in the hands of the eldest son, an obligation which continues to be observed to the present day. Today's collections in Wolfegg Castle consist mainly of the acquisitions of Maximilian Willibald.

The Housebook appears in the 1677 catalogue of the library of the fideicommission under "manuscripta": "manuscriptum chimicum on parchment, of Saturn etc. old" (Archive of the Fürsten zu Waldburg Wolfegg, no. 2899).

Nothing is known about the circumstances of its purchase, not to mention the seller. Two former owners have, however, left their mark in the codex itself: "This book belongs to Joachim Hof" (fig. 4) and, lower down on the same page, "lu[dwi]g Hof the Younger moved to

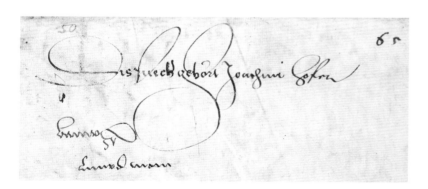

4 *Inscription of an owner's name on fol. 65r*

Innsbruck" (fol. 65v). Unfortunately, the names are too common to enable a link with any one individual. Equally insufficient for any identification are the hints at the original patron and owner, whose coat of arms appears twice in the Housebook: in full color at the beginning of the codex (fig. 1) and in an uncolored drawing in the middle (fig. 43). These arms are not of the old, simple geometric kind, divided into plain fields. The crest depicts none of the charges customary in heraldry, such as eagles or lions, but a golden treetrunk with sawn-off branches on a blue background. This is possibly a punning or 'canting' device, a rebus transposing the family name into an image, such as the later charge of the counts of Henneberg, showing a hen standing on a mountain. In the Housebook's case, there is a wide range of possible interpretations: Klotz, Stange, Buchner – the list of associations is endless and the search for a family has been correspondingly fruitless. That the family was probably bourgeois is indicated by the 'Stechhelm' (a type of helmet used for jousting), customary on commoners' heraldry, originally silver and now tarnished and black.

Nor is there any basis on which to date the codex. In the section on military affairs one page depicts a view of a large army encampment (fig. 61). Some of the heraldic shields shown here can be identified: the pacing lion on a blue background to the left of the emperor is the device of the counts of Nassau. It appears in a more impressive combination on the adjacent shield with a wheel, the arms of Mainz (fig. 5). The tent must therefore be that of Archbishop Adolf II of Nassau, who was both Prince Elector of Mainz and Imperial Chancellor. He died in Eltville in 1475. However, the artist who drew the shields made a few mistakes. Thus the coloring often changes from one representation of a shield to the next. A chevron charge, for instance, is shown once in red and silver, once in red and gold. The combination of the wheel and the lion on the Nassau shield is particularly unusual. The arms of the archbishops and prince bishops of Würzburg, Bamberg, Eichstätt, Mainz, etc. are normally clearly divided. In the case of Mainz, for instance, the wheel would appear in one field and the family arms of the holder of the appointment in the other. There happened to be families with coats of arms showing a lion holding a wheel, so the combination here could have led to a confusion everyone carefully sought to avoid. Either the artist was not conscientious enough or he could not remember the correct forms and colors.

A more concrete hint at a date is provided in the postscript to a prescription: "This is the Duke of Lorraine's piece, when nothing else was of help to him" (fol. 29v). It refers to the illness of René of Lorraine (1437-1508) in the winter of 1481-82, or more precisely to his recovery. This means that the prescription was written down after 1482.

5   *The banner of Adolf II of Nassau as Archbishop of Mainz and the arms of the patron, detail from the army camp, fol. 53r1*

The Medieval Housebook has frequently been the subject of art historical research. Just after the cultural and historical study by Retberg quoted from above, the Germanisches Nationalmuseum in Nuremberg published a volume of plates with engraved reproductions in 1866.[5] These two publications were complementary; a joint edition had originally been planned, but was abandoned due to disagreement between Retberg and the director of the museum.

A decisive step was taken in 1912 when Helmuth T. Bossert and Willy F. Storck published a facsimile edition of almost all the pages of the Housebook.[6] On this occasion the manuscript in its plain chestnut-brown leather binding was subjected to minute examination. Faded script was maltreated with reagents, which unfortunately left ugly bluish stains. The authors brought a wealth of details to light. Their bibliography listed over 200 items that had already dealt with the Housebook and its artist by then.

A slim popularizing volume with a nationalist element followed in the 1930s. Twenty years later a monograph by Johannes Graf zu Waldburg Wolfegg provided an entertaining description of the world of the Housebook.[7]

It was not until 1985, on the occasion of the centenary of the Rijksmuseum in Amsterdam, that a comprehensive exhibition was held of the oeuvre of the Housebook Master, or Master of the Amsterdam Cabinet. It was organized at great expense and also shown in Frankfurt at the Städelsches Kunstinstitut. The exhibition catalogue is one of the major publications on the subject.[8]

The first complete color facsimile edition was published parallel to the present volume. The accompanying commentary volume contains contributions by scholars specialized in various fields.[9]

The Housebook is a book. This sounds banal but should be emphasized. Scholarly research has treated it so far as an assemblage of master drawings in a more or less arbitrary sequence. Each drawing was thus evaluated and classified separately according to its artistic value, without reference to its context in the book. The result was a gradual splitting up of the artist's oeuvre. This may seem exaggerated, but interpretation has clearly tended to go in this direction. Ultimately, we can only do justice to the Housebook if we consider it a coherent whole – of both text and image – as we attempt to understand the aim and purpose of the important source material that the manuscript represents.

What are the relationships between this miscellany's various components, and is its composition even reconstructable in the first place?

THE COMPOSITION
OF THE MANUSCRIPT

In its present state the manuscript consists of 63 parchment leaves of uniform quality, divided into nine gatherings, each usually of four superimposed sheets folded in half and sewn together down the middle, forming eight leaves. They are enclosed in a parchment sheet, of which the front half was replaced in the 1985 restoration. The binding of plain flexible leather with flap looks like a folder.

Several foliations were made. One series of page numbers probably dates from the second quarter of the 16th century. It is in the same hand as the inscription stating Joachim Hof's ownership quoted above, below which "All / mine all" is written upside down and, lower down, "Of / total leaves / 66 [corrected to 65] leaves" (fol. 65r). Of special note is the numbering of the gatherings which was made earlier. They

6   *The Medieval Housebook*

are numbered from one to twelve, of which three, four and eight are now missing, indicating that there used to be three more gatherings in the past, that is, 96 leaves in all. According to this numbering, the Housebook originally consisted of twelve gatherings of four sheets (or eight leaves) each, provided reconstructing it with a uniform number of leaves per gathering is correct.

The codex today is not only missing gatherings but single leaves have also been removed: the very first leaf of the manuscript, six more in the chapter on the technology of warfare, and two most probably blank ones after the Hof foliation ends.

The folder-like binding is unusual for such a precious work of art. Such plain covers normally served to keep sheets together temporarily, until they were more appropriately bound. Similar provisional bindings have survived on archival items such as account books and collections of drawings where the binding was immaterial.

During the 1985 restoration the binding was carefully examined but with inconclusive results. The numerous different prickings and stitching clarify but little and are open to any interpretation. Some of the prickings could date from when the loose and as yet unwritten leaves were basted together in gatherings to keep them together. Which assemblage was made when is no longer discernible.

Whether we can conclude that the composition of this miscellany was constantly changing is not certain. While books with religious subject matter or secular narratives are commonly an integrated whole, it is typical for others, such as illustrated technical manuscripts, to be open-ended and expandable. It is therefore not unusual for a book such as ours to have blank pages and provisional stitching.

The only conclusion that can be drawn from the binding is that the Housebook was never part of a large library, where it would have been provided with a suitable cover.

The texts in the Housebook were written by two scribes. The first two gatherings, containing the art of memory and the verses on the planets, are by an experienced scribe in uniformly precise minuscules. The illuminated initials of the planet poems also exhibit the routine professional skill of a workshop. In contrast, the remaining texts in the Housebook are in cursive script. It is conscientiously tidy but not professional, its strokes sometimes firm and sometimes weak. The lines are often rather crooked and the scribe placed one opening initial so awkwardly that he had to lower the next line in order to avoid colliding with the initial (fol. 40r, third paragraph). Slight variations in script between the different gatherings suggest that the texts were not written at one go, but that longer periods of time elapsed between them.

The language points to South Germany and includes a surprisingly large number of Italian words. For example, one prescription states: "non cura de chi mangare." Such Italian expressions had become familiar just north of the Alps through the widespread trade relations maintained with Italy, especially Venice. Moreover, Italian terms were used in the business world in the way English ones are today.

While nothing in the manuscript as it is preserved today suggests what might have been on the third and fourth missing gatherings, that is, after the children of the planets, the fact that the metallurgical recipes in the ninth gathering begin with recipe d) indicates that they must have been preceded by at least three recipes in the missing eighth gathering. The six missing leaves after the mining section (fol. 39) probably contained material on war machines, as does the surviving leaf.

If the manuscript is reconstructed according to the given facts established as early as 1912, its structure turns out to be more carefully planned than it has seemed in the past. The first question that arises is why does the same full-page coat of arms appear twice in the same book. It must be more than a mere indication of ownership because one coat of arms prefacing the book would have sufficed.

The first coat of arms is at the beginning of the reconstructed codex, the second exactly in the middle. The Housebook is thus divided into two almost equal halves, each prefaced by a heraldic folio. Their function becomes clear in the second coat of arms. Oddly enough, it is surmounted by a 'sallet' ('Schaller'), a streamlined helmet not normally used in heraldry (fig. 43). Moreover, it looks as though it turned out too small in proportion to the shield and the eagle crest. This could be of some significance. In fact, the helmet depicted, equipped with a 'bevor' to protect the chin and neck, commonly served as a combat helmet, forming part of a man's equipment for battle. It is war that is the subject of the following chapter. Thus the coat of arms introduces a new section which contains with but one exception all of the material on technical constructions.

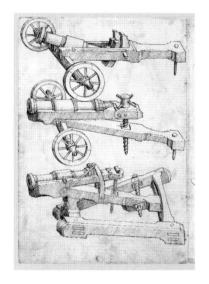

7 *Three cannons with various movable fixtures, fol. 56v*

The Subject Matter

The first part of the Housebook is the more difficult part to reconstruct because there is no indication of what might have been written or depicted on the missing two gatherings. The brightly-colored folio of acrobats or artistes forms the title page of a distinct chapter (fig. 9). It is framed in gold and positioned close to the fold, like a miniature taking the layout of the open book into consideration.

The artistes, dressed in red and green, perform in a broad hilly landscape with gently rolling hills crowned by fortresses on the horizon. If the acrobats were removed it would be a fitting backdrop for St. George killing the dragon.

In the foreground, two men with practice swords are giving a performance, separated by an umpire. Crossed pairs of staffs, daggers and Turkish sabres lie on the ground. Above, two artistes grasp each other by the hand, about to start wrestling. A fire-swallower and a snake charmer display their skills next to an artiste who, despite the blood dripping from his wrist, appears to feel no pain from the dagger thrust through it. In the background on the right, a few spectators are gathered around a distinguished man. This small group derives from a copperplate engraving of *The Martyrdom of St. Sebastian* by Master E. S. (fig. 8). He was active in the Upper Rhine area a generation before the Housebook Master, between 1450 and 1467.[10] The Housebook acrobats themselves are also borrowed from the work of Master E. S. They appear in the larger of his four decks of playing cards, dating from 1463. It has coats of arms, dogs, birds and a wide variety of human figures for its four suits (fig. 10).[11]

8 *Detail from an engraving of 'The Martyrdom of St. Sebastian' by Master E.S.*

In an interplay of mutual influences which can scarcely be reconstructed today, motifs for such playing cards were usually taken from model books, while conversely the cards themselves were often favorite models for artists. This stock of motifs was especially popular as a source for decorating the borders of luxury manuscripts, whose gambolling figures entertained their readers.

Possibly the people in the major card deck of Master E. S. were themselves modified borrowings from the figures in the illustrated fencing guide by the fencing master Hans Thalhofer, of which they are strongly reminiscent.[12] It was not uncommon for war manuals to have recourse to the teachings of fencing and wrestling. This would explain how a standard component of such codices found its way via a detour

9 *Introductory miniature of artistes or acrobats, fol. 3r*

into the Wolfegg manuscript. The artistes betray their origins. Here too, they are arrayed as though against the flat background of a playing card. In rows one above the other, they are distributed evenly across the page without any contact with each other or relationship to the landscape.

The castle buildings on the right and the little rider curiously galloping up hill and down dale derive from another work by Master E.S., *The Large Garden of Love* (fig. 37).[13] This engraving has been virtually exploited by the Housebook. The main scene, a licentious company of diners, was actually taken over in its entirety in another drawing. The large castle on the left hill in the picture of the artistes also belongs to the repertoire of Master E.S. It appears in his *Nativity* (fig. 11).[14]

The miniature has an enumerative and indicatory character. Its individual motifs are placed side by side in a representative way without any convincing relationship to each other. Despite its spatial inadequacies it was carefully framed in gold. The miniature is puzzling. It was obviously intended as a title page, but one can only speculate

10 *People suit – four, of the 'Large Deck of Playing Cards' by Master E.S., 1463*

11 *'Nativity,' engraving by Master E.S.*

about its literal meaning. The small group at the upper right could stand for medieval society, with a ruler in the middle flanked by a young knight and a scholar in a long gown. The artistes are divided into two groups: one displays fairground variety acts, another wrestling exercises. Each performs on a hill of its own. There seems to have been a reason behind this. The two hilltop castles that do not disappear into the blue haze like the others are completely different: the right castle is closed off and well fortified with firing slits and crenallation. No entrance is in sight; the massive walls are forbidding. The steep hill on which it was built grows only a few scattered trees, as an unobstructed view and a clear field of fire are indispensable for defense. A gray horseman is leaping towards the other castle, his arm raised high. The sun shines brightly on this one, more of a palace than a fortress. Its wide open gate is in plain view; its gilt rooftops sparkle in the sunlight. This building is more refined and detailed. The slope gently rising on one side is wooded and by no means as unwelcoming as its counterpart. Perhaps the groups of figures could also be distributed correspondingly. The fighting group would hence belong to the fortified castle, the trio with their strange stunts to the ornate palace. That the conjurers come from a more cheerful environment is also indicated by the gold embroidery on their doublets.

In the final analysis, the significance of these contrapositions remains obscure. The miniature has a counterpart in the mining panorama (fol. 35r) introducing the second part of the manuscript that contains drawings and texts on mining. Perhaps a corresponding connection could be made between the artistic skill of the artistes and the arts that follow. In this case they at least provide an impressive prelude as the title page of the chapters on the liberal arts, with the art of memory representing rhetoric, and the deities of the planets astrology. All of this must remain speculation, for the last word on the construction of the Housebook has not been spoken yet.

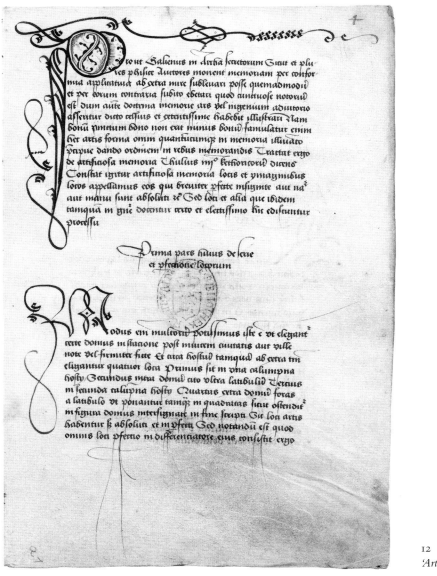

12   *The first page of the
'Art of Memory,' fol. 4r*

THE LIBERAL ARTS

The most important arts in the medieval educational system were the 'artes liberales,' originally those deemed worthy of a freeborn citizen. Ever since late antiquity these consisted of seven arts subdivided into two groups: the 'trivium' composed of grammar, rhetoric and dialectic, and the 'quadrivium' of the mathematical sciences of arithmetic, music, geometry and astronomy.

Other arts coexisted with these, such as the lesser 'artes mechanicae,' which included activities with which to make a living, such as the crafts, as well as warfare. There was also a series of forbidden arts, including everything pertaining to magic and divination, such as geomancy, palmistry, or the strange art of onomancy, the art of predicting the future from a name. The latter appears in many a codex on the technology of war. In the course of the Middle Ages the number of forbidden arts was adapted to the scheme of seven to correspond to the liberal and mechanical arts.

The Art of Memory

The art of memory had from time immemorial been regarded as an auxiliary of rhetoric. It was to help the orator remember certain key points of his speech. The 'ars memorandi' also served the general purpose of memorization and contemplation, for example in order to be able to recall the sequence of the chapters of the Gospels from memory. It provides various methods which include associating concepts with simple images – a snail, for instance, stands for 'slow' – as well as more complex systems requiring more practice. Such a system is presented in the Housebook in very difficult Latin.[15] The treatise is only three pages in length and describes a method transmitted in many other manuscripts. It is based on imagined places: one imagines a house with people in certain places within, doing things associated with the material to be remembered, the more nonsensical the better, because it will be easier to remember. Space was left free for a drawing of the house referred to in the text, but it was not carried out. The text in the Housebook is often unintelligible or simply incorrect, suggesting it is the work of an adapter who was not experienced with the terminology of the genre. Or perhaps the scribe presumed the reader to be so familiar with the art of memory that he only had to record major points in keyword form.

The art of memory has been placed in the most prominent position in the Housebook, namely at the beginning of the manuscript. This implies that it has a key function for the whole book. However, we can only speculate about its point – as about that of the artistes.

After the art of memory follow the planets in their customary sequence. Astrology was still closely linked to astronomy at the time, an association which only gradually dissolved, finally liberating science from the obligation to integrate every newly discovered phenomenon into a coherent cosmology. The pictures of the planets and their influences in the Housebook do not presuppose any complicated mathematical calculations. They are not very far from the horoscopes in today's newspapers: astrology for the man in the street in search of an instant pat answer.

In ancient tradition according to the Ptolemeic system, the planets were disposed in order of their orbital periods, with the earth at their center. It was an integrated system in which every part had its place and was subject to divine power. The earth was thought to be surrounded by layers of invisible spheres onto which the planets were attached. These spheres slowly revolved inside each other, causing the planets to move, with the North Star as their fixed point.[16]

Someone looking at the heavens would notice a strange phenomenon after a while: some of the stars move through the constellations. Hence the ancients called them planets, from the word meaning wandering, or roaming. They were familiar with five planets: Mercury, Venus, Mars, Jupiter and Saturn. (The other three in the solar system known today, Uranus, Neptune and Pluto, were not identified until between the 18th and 20th centuries with the help of telescopes.) The five planets are usually brighter than the fixed stars and their courses run close to the annual course of the sun, the ecliptic. The sun and the moon also used to count as planets because they look like they move.

The individual characteristics of the planets evolved in conjunction with the theory of the four humors. The Hippocratic writings on the humors correlated the four elements (fire, earth, water and air) with the four humors operating in man and with the corresponding temperaments. Thus yellow gall stands for fire, and black gall for the cold, dry earth. Phlegm is equated with water, and blood with air. Because the human body is determined by these four humors, these interrelations are significant. If yellow gall predominates in someone, he is sure to have the explosive and spiteful temperament of the choleric, because fire or yellow gall are determined by their dry and hot qualities. On the other hand, water is cold and wet, as is phlegm, and so are the phlegmatic people, dozing away, lazy and bloated. Black gall with its dry and cold consistency is accompanied by depression and frigidity. The person dominated by blood, the sanguine type, is fortunate – with a cheerful and serene temperament.

The Medieval Housebook

Facsimile and Commentary

Prestel

This famous work of medieval drawing offers a unique and vivid insight into the way of life and thinking in the world 500 years ago

Incomparable in its vivid portrayal, the famous 'Medieval Housebook' provides a humorous insight into the world of the late Middle Ages in Germany. In addition to a discourse on mnemonics, a section on astrology with drawings of the planets and on the life of the knights — complete with tournaments and gardens of love — medicinal and household recipes can also be found as well

The collection

The foundation of this important collection belonging to the Princes of Waldburg Wolfegg was laid in the mid-17th century by the Lord High Steward Maximilian Willibald (1604-1667). In contrast to the majority of important royal collections of engravings and drawings, the Wolfegg cabinet did not pass into a public museum but has survived to this day in its traditional home in the print room at Schloss Wolfegg.

In front of a castle surrounded by water — day-to-day life goes on, with duck hunting and fishing (fol. 19v-20r)

Preparations for 'crown stabbing', a tournament in which the opponents try to knock one another from their horses with lances (fol. 20v-21r)

as chapters on mining and the art of warfare. Thus the manuscript sketches a wide-ranging picture of thought and knowledge at the time. The Housebook, completed in 1480, is particularly important due to its artist, who portrayed his world with wit and sensitivity. The so-called Housebook Master, who worked in the Middle Rhine region and is known from several other works, had considerable influence, particularly through his engraving, on his younger contemporary, Albrecht Dürer.

The identity of the artist as well as the patron remain one of the greatest mysteries of art history.

Mining panorama

For over 300 years, the manuscript has formed part of the collection of the Princes of Waldburg Wolfegg at Schloss Wolfegg near Ravensburg.

The contents of the 'Medieval Housebook'

Artists
The 'artes liberales':
  Mnemonics
  Drawings of planets
The life of the knights:
  Bathhouse and
    moated castle
  Tournaments
  Hunting and love
  The obscene garden
    of love
Household remedies
Mining, smelting and
  minting
The art of warfare

Troubadours and artists

The original of the 'Medieval Housebook' will be on display in the following locations:

Städelsches Kunstinstitut, Frankfurt a. M. (18.9.-2.11.1997)
Rijksmuseum Amsterdam (22.11.1997-18.1.1998)
Haus der Kunst, Munich (24.7.-11.10.1998)
National Gallery, Washington (15.11.1998-31.1.1999)
The Frick Collection, New York (11.5.-25.7.1999)

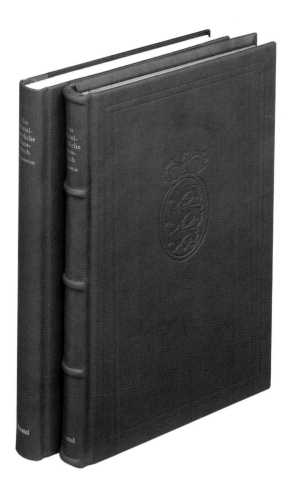

The 'Medieval Housebook' as a facsimile with a commentary volume in a strictly limited edition

The Medieval Housebook
Das Mittelalterliche Hausbuch

Limited facsimile edition with commentary volume in 750 numbered copies.

Edited by Christoph Count of Waldburg Wolfegg.
With contributions by Gundolf Keil, Eberhard König, Rainer Leng, Karl-Heinz Ludwig and Christoph Count of Waldburg Wolfegg.
Format 7 ¾ x 11 ¾ in. ISBN 3-7913-1838-1.
The volumes are only available together in an ornamental slipcase.

To give an impression of the high quality of the edition, a file is available containing three pages reproduced in facsimile in the original format, together with a detailed description of the facsimile for a token fee of US $ 65.00. When buying the facsimile edition, the price of this file will be deducted.

The edition:
The two-volume publication will appear as a facsimile with a commentary volume in a strictly limited edition. Volumes 1–750 will be numbered by hand in Arabic figures and intended for sale. In addition to these, there will be 50 volumes numbered in Roman numerals which will not be available on the market. When printing is complete, the printing plates will be destroyed, thus guaranteeing the uniqueness of the edition.

Facsimile volume: To reproduce the facsimile, the original manuscript from Schloss Wolfegg has been taken apart by experts. The 130 pages have been reproduced with the utmost care and printed in five colors (with gold). The printed sheets are cut individually and bound by hand. The full leather binding is embossed with gold on the spine and, on the front cover, with the coat of arms of the Princes of Waldburg.

Commentary volume: The accompanying book is published bilingually in German and English. Approx. 236 pages and 48 b/w illustrations. Half-leather binding with embossed gold on the spine.

The first colored 'Housebook' facsimile: a must for all collectors and lovers of book illumination

◄ In the 'Medieval Housebook', seven planets appear at the beginning, arranged according to their periods of revolution. These affect the character of the individual at the hour he is born. The last planet with the fastest period of revolution is the moon. The children of Lady Luna are thus predestined for careers linked with water, such as fishing or milling, or something suiting their changeable and freedom-loving natures, e.g. bird-catchers, magicians or artists.

"The stars have worked their magic on me, I am unsettled and strange. My child can barely be tamed, she will be subject to no one... And he who makes his living from water, shall be blessed by the light of the moon."

Following the tour of international exhibitions from 1997 to 1999, the priceless original of the 'Medieval Housebook' will be reassembled and newly bound, to be locked in the safe at Schloss Wolfegg.

The famous manuscript will subsequently only be accessible in the form of this first colored facsimile. The edition will thus become a very sought-after collector's piece and an invaluable source for research.

Prestel   Mandlstraße 26   80802 Munich   Phone 089/3817090   Fax 089/335175

✂ -

Order Form

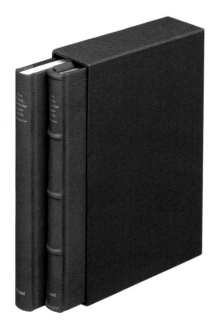

☐ Yes, I would like to order the facsimile edition of the 'Medieval Housebook' with commentary at the price of **US $ 1,980.00**
(Prestel will allocate the edition numbers chronologically as orders are received)

☐ Yes, I would like to order the documentation file for the 'Medieval Housebook' for a token fee of US $ 65.00
(When buying the facsimile edition, the price of this file will be deducted)

Name

Address   City

Country   Phone (in the event of any queries)

Date   Signature

Please complete and send to your local bookshop

or mail to

Prestel – Verlag
Mandlstraße 26
D 80802 München / Germany
Fax: +49 / 89 / 335175

ISBN 3-7913-9967-9, 5 Pera 798

In Arabic thought, according to a theory current by the 9th century, the four humors and the planets were aligned with colors: Saturn, which appears dark, correlates to the color black and to cold, dry black gall. Red Mars corresponds to yellow gall, Jupiter to blood, and pale Luna to phlegm. The astronomical and physical qualities of the planets were linked to those of the ancient Olympian gods, who thus survived into the modern age even outside academic circles. Others who lacked a planet, such as Pallas Athena, missed out on such popularity. In the medieval West, despite opposition from the Church, astrological teachings interwove with Christian ones. Alexander Neckam, among others, polemicized against the influence of the planets on human fate,

13 *The four temperaments: sanguine, choleric, phlegmatic and melancholy, woodcuts from the 'Augsburg Calendar,' c. 1480*

14   *The planet Jupiter,*
detail, fol. 12r

but noted: "Seven are the planets that not only adorn the world but also exercise their influences on the lower sphere, influences that were conferred upon them by nature in the highest, which is God."[17]

To study the planets was natural for the scholars, if only to be able to interpret ancient sources. Some gods, such as Saturn, underwent a transformation. In the course of the Middle Ages he lost his positive qualities, and eventually stood only for poverty and misfortune. Finally, late medieval popular abbreviated versions greatly simplified the earlier complex intellectual edifices by merely enumerating the characteristics of the planets in key words.

From a geocentric point of view, the planets all move through the zodiac. The relationships of the planets to the signs of the zodiac are therefore of considerable importance. Depending on what sign a planet is moving through, the planet's power can be increased, reduced, or even eradicated. Every planet rules one or two signs of the zodiac, which generally have similar qualities. Jupiter is thus the ruler of Sagittarius and Pisces. They are shown in the Housebook next to the mounted planetary god (fig. 14).

The medical context implied above becomes more evident in the bloodletting treatises which were extremely popular, especially in the

Late Middle Ages and early Renaissance. As opposed to serious works, however, the texts on the planets in the Housebook are intended for domestic use and they simplify their complicated models substantially.

Science today differs from that of the Middle Ages. We are used to a high degree of specialization that sees nothing beyond its own field. In those days, however, thinkers always tried to invent models that were comprehensive. The human being was considered a whole. An imbalance of the humors was thought to be the cause of disease, which in this case meant that balance and order had to be restored. The observation of the skies was approached in the same way: over everything stood heaven and the Almighty. Order determined the way of thinking. Disorder in the upper spheres near God meant disorder and chaos down on the earth far from the divine. A meteor was therefore interpreted as a sign of something extraordinary, mostly negative, with consequences for human beings. Such a holistic view had its advantages – quite obvious in medicine – as well as a grave disadvantage: every innovation had to be brought into harmony with the entire system in order not to threaten it. The observation of a new apparition in the sky in 1577, presumably a supernova, caused the ancient dogma that the heavens were unchangeable to collapse. Ever more serious contradictions arose, especially in astronomy. Eventually, as the centuries went by, the need to explain anything and everything by means of a system revolving around man was gradually abandoned.

Astrology is basically that part of ancient astronomy which continues to preserve the old complexity of the relationship between the universe and man. It perpetuates the idea that the cosmos was created for the pleasure of the human beings who dwell at its center. As a halfway reputable science, astrology survived into the 17th century – Kepler had 800 horoscopes in his manuscripts – and only gradually lost its respectability later.

The age of the Housebook was still an era that believed in astrology. But as far as the contents of the verses in it are concerned, they are on the same level as today's newspaper horoscopes, in other words, aimed at the general public.

Saturn

The series of planets begins with Saturn, who takes longest to cross the heavens, namely about thirty years (fols 10v-11r). He is therefore often characterized as an old man with a crutch. As do all the other planetary deities depicted in the Housebook, he has the attributes of a jousting knight. He carries a lance and his horse's caparison is ornamented with bells that serve to distract the animal from unusual noises. A hissing dragon decorates his pennon. It is the dragon that swallows time, for the ancient deity of Saturn was amalgamated with Chronos, the Greek god of time. Capricorn and Aquarius are his signs of the zodiac. The accompanying verses are divided into two parts. The first describes the planet itself:[18]

> "Saturn is my name; I'm first
> of planets high above the earth.
> I am by nature dry and cold,
> and my works are manifold.
> I in my houses firmly stand:
> the Goat and the Waterman.
> I do much damage by my might
> on sea and land, by day and night.
> My exaltation's in the Scales,
> but in the Ram my power fails.
> It's thirty years, harsh and malign,
> ere I come again to the same sign."

The second part describes the appearance and qualities of the children of this planet, who are the people born under this sign:

> "My children are vicious, dry and old,
> envious, weary, wretched, cold. [...]
> They grub the dirt, dig graves, plow land,
> in foul and stinking clothes they stand. [...]
> always needy, never free [...]."

The peasant in the field and the laborer digging an irrigation ditch with a pick and shovel both have a hard life, as does the grotesque renderer, who loses his trousers and is sniffed at by a sow. Such work is actually one of the more pleasant aspects of Saturn. As Saturn's child you could end up in the stocks or, at worst, finish your life of crime on the gallows or on the wheel, your broken limbs picked at by crows. One villain with such a fate is shown. In torn rags, bound, and attended by a priest, he is led to the gallows hill by a hangman.

15   *Saturn and his children, fol. 11r*

Jupiter

The next planet and his children promise more cheer: it is Jupiter, the father of the gods, accompanied by the signs of the zodiac of Sagittarius and Pisces (fols 11v-12r). On his children:

> "Fortune smiles, they're just and wise,
> rosy faces and laughing eyes.
> Well-mannered and well-clothed, refined,
> with hound and bow they hunt the hind.
> In falconry they have much art.
> Well-mounted, they pursue the hart."

These are the pastimes of the aristocracy that are enumerated here: stag hunting and falconry.

In a fenced-off area, three archers are target-shooting with cross-bows. The strange crank used by one of the marksmen is a tooth rack winch. This device facilitated drawing the arbalest, which took a lot of strength. The marksman with his back to us has a simpler model. He presumably draws back the bowstring with a hook on his belt. Next to the roofed target, an aide seated behind a protective wall shows the score.

As another typical aristocratic pasttime, the adjacent scene shows the classic stag hunt with hounds to chase the game. An elegant couple on a leaping horse also has a falcon. Small birds such as quails or partridges could be hunted with hawks. Larger ones, such as cranes or herons required a falcon. Falconry called for certain equipment, including the glove clearly identifiable here, into which the bird could dig its claws. The narrow leather straps always around the falcon's feet are also visible, dangling down. We will encounter the falcon as a young man's companion elsewhere in the Housebook.

Several small buildings in the foreground – based on similar structures in illuminated manuscripts – house further children of Jupiter. They are judges, jurists and scholars. A pair of pointed clogs lies on the floor next to the scholar's feet on the left. These were slipped on over elegant pointed shoes to protect them from the mud outdoors. The judge enthroned under a baldachin wears them too. In front of him, a poor man supported by a companion awaits his judgement, while a portly man opposite speaks insistently to the judge with emphatic gestures. His fleshy face and fur-lined clothing proclaim that he is one of life's priviledged few.

16 Jupiter and his children, fol. 12r

Mars

The third planet, with a course of about two years, is warlike Mars (fols 12v-13r). He dashes from left to right, as did the unpropitious Saturn, which does not bode well. The planetary horseman wears armour. The besagews (discs) at his armpits are meant to prevent the enemy's lance from hooking in. Even the steed's shaffron (head defense) is unpleasantly prickly. The scene below does not depict a battle in which equally armed adversaries oppose each other. Instead, the evil children of Mars have sought out weak and defenseless victims. One of these is the pilgrim in the foreground, with the typical accoutrements of a staff, shoulder bag, rosary and felt hat decorated with pilgrim's badges. These are little plaques sold at pilgrimage sites. Most familar was the scallop shell from Santiago de Compostela, sold in all sizes and in great numbers in front of the cathedral. The peaceable pilgrim will never reach his goal. A robber has knocked him down and stabs him with a dagger.

The building with a stepped gable on the right is a shop. The opened shutter was let down to form the counter. Thieves seize the opportunity to steal everything on it while the shopkeeper is being murdered indoors. The other acts of violence on this page are perpetuated by soldiers. They set fire to a village and misappropriate the cattle. Two women defend themselves by thrashing the soldiers with a jug and a spindle, respectively. Most of the villagers hide in the only fortified building in town, the church. Their tiny frightened faces peer out the windows.

17   *Mars and his children,*
fol. 13r

Sol

After the gruesome events under Mars, comes Sol with the sign of Leo (fols 13v–14r). The regal sun has a positive effect on people born during its reign. They are pious and enjoy life. After church they give alms to the beggar crouching on the floor. Then the merry children of Sol meet for a snack while listening to a concert played by a trombone and wind ensemble. Beyond the garden wall, across which a fool is trying to disturb a pair of lovers, young men disport themselves in physical exercise: stone-putting, fencing and wrestling. In the happy world of bearded Sol, the falcon must not fail.

"Men call me Sol, I am the sun,
the middle planet, on I run.
Beneficient and warm and dry
by nature, my rays fill the sky.
The Lion's in my house, therein I dwell,
and brightly shining I do well.
There I stand, fair and bold,
against old Saturn's bitter cold.
In the Ram I rule and reign,
but in the Maid I fail, I wane.
And through the stars my way to wend,
three hundred and sixty-five days I spend.

Noble and fortunate I am,
as are all my children. [. . .]
Happy, kindly, well-born, strong,
fond of harps, viols and song.
All morning long to God they pray,
and after noon they laugh and play.
They wrestle and they fence with swords,
they throw big stones, and serve great lords.
Manly exercises are their sports,
they have good luck in princely courts."

Venus

The next planet is Venus. Strangely enough, she turns her back to us as she rides by, accompanied by Libra and Taurus (fols 14v-15r). The distinguishing qualities of the people born under her influence are illustrated on the ground below.

At the lower right, things are very courtly and proper. A slow dance of elegantly dressed couples is accompanied by a brass trio. In contrast, the goings-on behind them are coarser. Here the instruments include the bagpipes, interpreted as an obscene allusion to male genitals in those days. The dancers are uninhibited, and a peasant couple tumbles in the nearby bushes. In a leafy arbor in the left foreground, with a real door and a wattle fence, a nude woman climbs very unchastely into a tub to join a waiting youth. An old procuress waits alongside with some food and wine.

"I am Venus, fifth planet above
the world, and am the light of love.
I'm moist and cold and in my hour
men feel my great and awesome power.
Two houses are mine, in which I fare:
the Bull and Scales. And when I'm there
I live in joy and jollity,
and Mars can never frighten me.
In the cold wet Fishes I'm glad to rise;
in the Virgin's sign my power dies.
In just one year and then one day
through all the signs I gently play.

Lightly loving, full of mirth,
my children are happy here on earth.
Merry when rich and merry poor,
none can compare, you may be sure.
Pipe and tabor, harps and lutes,
they play organs, horns and flutes.
With singing and with dancing too,
embrace their lovers, kiss and woo.
They rejoice to hear fair music's sound.
Their mouths are darling, faces round.
Beautiful bodies, parched by lust's heat,
my children find love's duties sweet."

19   *Venus and her children,*
fol. 15r

Mercury

By contrast, the next planet with its signs of Gemini and Virgo is rather dull. Mercury's children are industrious. Besides craftsmen, such as goldsmiths, painters, sculptors and organ builders, they include teachers (fols 15v-16r). This is the only scene showing the children of the planets indoors. A pair of spectacles perches on the nose of the goldsmith in his workshop in the left foreground. While he fashions a beaker, his wife has to pump the bellows. This does not stop her from watching her husband at work over her shoulder. A teacher's customary attribute was a switch. One is vigorously being made use of in the classroom here. Beside it, an organ builder tunes the pipes while his apprentice sullenly operates the bellows.

Near the right margin, the artist's wife lays her hand familiarly on the shoulder of her painting husband. She watches an altarpiece develop, which depicts the Madonna and Child with St. Catherine and her wheel – as we do. The sculptor in the foreground also has support at work: he is being handed a beaker by a lady at a table set for a meal. A dog under it waits eagerly for a treat.

Luna

The quality of the drawings changes from one to the next. While Saturn was somewhat coarse, Luna, as the last in the series, is of consummate refinement and beauty (fig. 21). The ink has been deployed in inimitable fashion to shape the mounted goddess. Sunlit surfaces are a delicate brown, while parts that are in shadow are often deep black. It is the most charming drawing in the Housebook.

The moon, quickest to circle the earth, engenders people who are difficult to govern and are vagabonds, such as conjurers. Its wet nature corresponds to millers and fishermen, among others.

"Cold and wet my power ranging
over all, unstable, changing. [...]

Headstrong, heedless and half wild –
if he won't be led, he's Luna's child.
Pale round faces and brown eyes,
cruel teeth, snub-nosed and never wise [...]
If you fish or swim or sail,
as Luna's child you cannot fail."

Luna der monat der letzst planet naß
heiß ich vnd wurtk dmyk die sem laß
kalt vnd feucht mein wurckung ist
Naturlich vnstet zu aller fast
Der kebs mein hawß besessen hat
So mein figur dor inne stat

Vnd Jupiter mich schawet an
kain vbels ich gewurcken kan
Erhohet werde ich in dem stir
Im storp valle ich nider stir
Die zwelff zeichen ich durch gang
In sibenvndzwentzig tagen lang

Der sterne wurcken geet durch mich
Ich pin vnstet vnd wunderlich
Mein kint man kaum gezemen kan
Nymant sein sie gerne vntertane
Ir antlutz ist plaiß vnd runt
Prawn grausam zene ein dicken munt
Vbersichtig schele einen entzen gemut
Sein hofferttig trotz der leib ist mit lanck
leuffer gauckler fischer marner
farn schuler vogler maler pader
Vnd was mit wasser sich ernert
Dem ist des monats sitten beschert

21   *Luna and her children*
with the accompanying text,
fols 16v–17r

40   *The Children of the Planets*

A fowler hides in his hut camouflaged with leaves in the right background. In order to attract birds, he suspends an owl on a stick on front of his hut. As many birds dislike owls, they will attack as soon as they see one. When a bird lands to do so, the fowler catches it with a pole with a clamp mechanism attached. The fowler's dubious reputation has long been a topos. He was said to satisfy abandoned wives on his wanderings.

The drawings of the planetary gods in the Housebook represent an odd mixture of different traditions. A group of handbooks on astrology and medicine contain the same repertoire of pictures as the Housebook. Besides manuscripts, they include some blockbooks that have been preserved and are very difficult to date, presumably from the 1460s (fig. 22).[19] I suspect that one of these blockbooks was the source for the Wolfegg codex. This would explain the awkwardness of the drawing style at the beginning of the series, although it surpasses the incunabula by far in quality. In the clumsy, artistically rather inept images of the blockbooks the main emphasis had been on naming all the children of the planets.

Moreover, there is a significant difference between the Housebook and the earlier astrological handbooks. The latter usually depict the planets standing up; even the warrior Mars is always unmounted. Riding deities, though without the children of the planets, can be found in Konrad Kyeser's manuscript on war technology, *Bellifortis* (fig. 23) and its copies, though his followers lack any reference to the planets. Whether Kyeser actually furnished the model for the Housebook cannot be ascertained, though the jousting armour they have in common speaks strongly for it. It remains an open question to what extent the combination of horseman and children of the planets is an invention of the Housebook Master.

The drawings of the planets in the Housebook reveal two different styles. The drawings of Saturn, Jupiter and, a few leaves later, Mercury feature vigorous, angular outlines and distinct forms. The bathing scene in the Venus picture is similar, differing from the rest of the drawing, where the other children of Venus are in a softer, more delicate style. A related, more painterly drawing style characterizes the pictures of Mars, Sol, and the last of the planets, Luna.

This has led to the attribution of different drawings to different artists. In fact, however, the explanation for the apparently abrupt changes in style can be found in book production. The sheets forming pairs of leaves were first illuminated before they were laid on top of each other to form the gathering. It turns out that the woodcut-like drawings of Saturn and Mercury are on one sheet, Jupiter and Venus on the next. Work was evidently interrupted when only a few of the

22   *Luna and her children,*
blockbook fragment,
Schwabach Church Library

23 *The planet Sol in a 'Bellifortis' manuscript, Göttingen, Staats- und Universitätsbibliothek, Cod. ms. philos. 63 Cim.*

children of Venus had been drawn, and continued at a later point. The differences in the drawings can thus be accounted for by the artistic development of the artist. He completed the series of the planets in his mature style with Mars, Sol and, on a separate leaf, Luna. Proving that work was indeed carried out before the sheets were bound, captions label the pictorial pages: "Saturnus," "iupiter," "Mars," etc. They constitute a helpful precaution because the text did not appear next to the picture on the unbound sheets. This sequence of production is corroborated by the increasing size of the planetary riders. While the first deity, Saturn, is suspended like a solitary star in the sky, the last one, mounted Luna, occupies almost half the scene.

Astrology interrelates the macro- and microcosm and classifies these under the different planets with their signs of the zodiac, as shown here. The very first text in the Housebook, the art of memory, constructs a mental building to retain what is worth remembering and to systematize it at the same time. It seems that there is more behind the Housebook's selection of texts than their conventional and popularizing subject matter would at first suggest.

The tone changes in the two gatherings that follow. They include the apparent genre scenes, followed by a bundle of leaves with prescriptions and recipes of various kinds, and finally the drawing of a spinning wheel.

After the drawings of the planets, two gatherings are missing from the manuscript. The next seven drawings, each covering two facing pages, depict the so-called genre scenes. They are probably the most famous pictures in the Housebook.

Neither the drawings of the planets before them in the book nor of the machines after them are in an arbitrary sequence. These scenes are therefore most probably also subject to a plan. The only question is, what is it?

24   *Spectators at the coronel joust, detail, fol. 21r*

The cheerful insouciance of these compositions is striking. They lack the warning raised finger of the moralizing Netherlandish art to come. Even in the most serious and dangerous scene, the warlike joust, tension is defused with humor: In a furious horse race sweeping across the page, one of the horses is seized by the tail and thus prevented from starting (fig. 32). This friendly dissimulating manner is utterly typical of the Housebook Master and can be perceived in many of his engravings. Various facets of this endearing world view can be observed in subtly differentiated nuances throughout the Housebook.

Bathhouse and Water-Surrounded Castle

In the first scene of the series, an elegant lady is introduced to a world of leisure (fig. 25). An airy bathhouse stands in a garden protected by a high wall. A couple entering the garden with another woman pauses while the lady in front glances up at a tall fountain. Its column is topped by a figure squirting water straight up into the air through a horn. Two other couples entertain themselves by reading aloud and walking. Meanwhile, a mischievous monkey menacingly dangles a ball and chain, intended to prevent him from running away, over a grayhound as yet sleeping peacefully below. The monkey and grayhound occur in virtually identical positions in the foliate ornament of a pontifical made for Adolf of Nassau, Archbishop of Mainz, who died in 1475. His tent appears in the encampment scene, as mentioned above.[20] A large arched window provides a view into the bathhouse, where both sexes are enjoying themselves. A lute player sits in the window; a youth leaning casually on the table in front of it listens idly to the music.

The coloring of this drawing is more restrained than that of the others. The parchment was even left blank on parts of the clothing by way of highlighting. On other pages the corresponding garment is usually completely filled in with paint. Gold was used very sparingly, mainly for buttons and single pieces of jewelry. The drawing's erotic flavor is unmistakable, which is underlined by its links with the Venus page.

The interpretation of this scene was the subject of scholarly debate between cultural historians who adduced a wide range of pictorial matter to support their theories.[21] Norbert Elias saw the Wolfegg codex as evidence for how a member of the upper classes at the time perceived his environment. "It simply tells the story of how a knight felt and saw the world," he writes, corroborating his theory of the progress of civilization away from the untroubled naïveté of the Middle Ages. Without realizing it, Elias has allowed himself be fooled by the Housebook Master. As in the representations of the planets and their children, what is apparently true to life is the vitality of a fictitious world. In contrast to the obviously stereotypical character of earlier woodcuts and blockbooks on the same subjects, the verisimilitude of the Housebook's drawings is deceptive. In their imaginary world, other things are possible than in reality, including some incongruities. By comparison, the reader of an illustrated edition of the *Roman de la Rose* would know that this is literature, and would never expect to find everyday life in the Middle Ages reflected in it. The case of the Housebook is more complicated because most of its contents do comprise practical material relating to reality. But the scenes in question, though they no doubt reflect contemporary notions, make it hard for us to determine how consistent they are with real life in that period. To return to the bathhouse, the drawing itself permits no conclusions about whether it was customary in those days for both sexes to bathe together. Morality was far stricter than Elias thought. Communal bathing in the nude was only possible in establishments that were like brothels. Does that mean the Housebook portrays a house of ill repute? Is it an, admittedly male

45

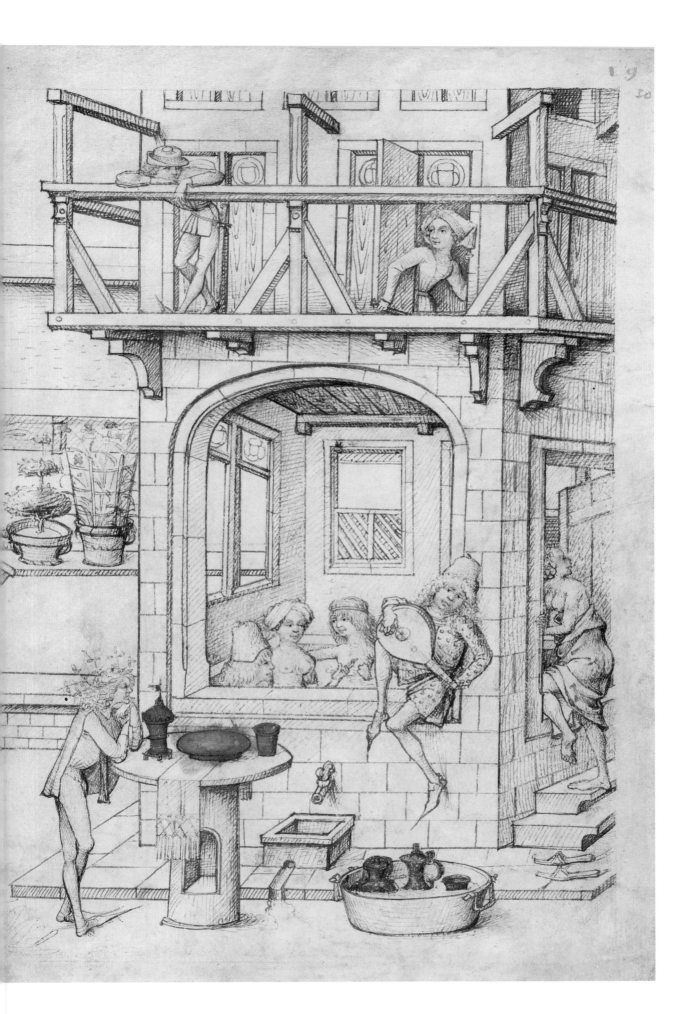

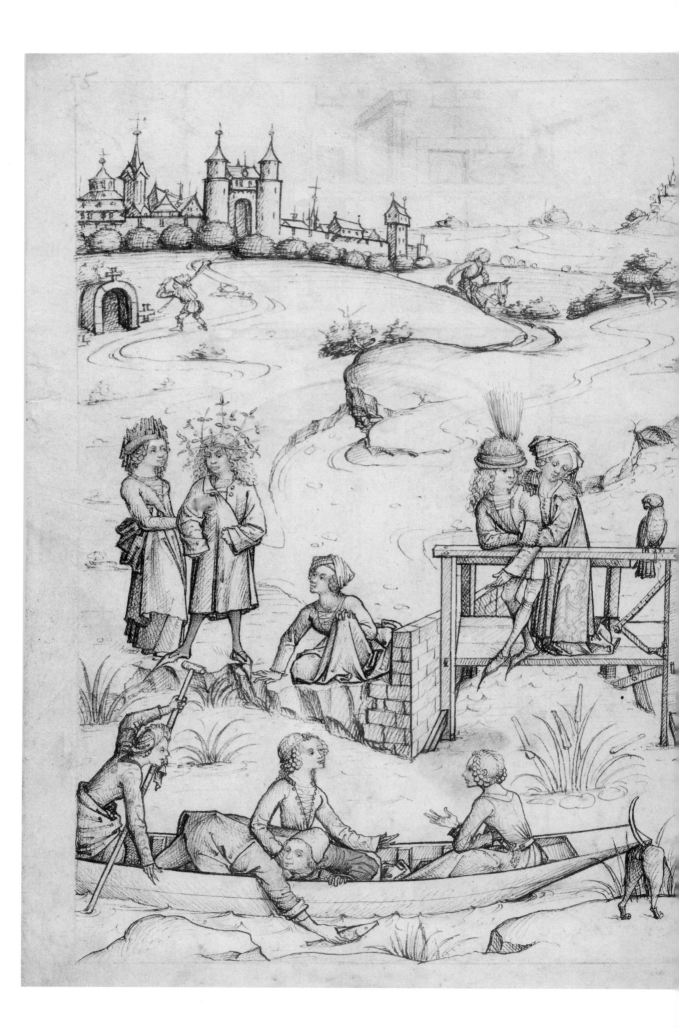

fantasy? Or are we dealing with a piece of fiction in which the nudity that was normally immoral is not considered offensive? The subsequent drawings yield further insight.

In the drawing of the water-surrounded castle that follows, water again plays a major role. Here it fills a moat surrounding a compact functional building with many towers and crenallations. It stands in an open landscape with a fortified town in the background (fig. 26). It is chilly, people are wearing coats. Similar to the youth in front of the bathhouse, another in a short cloak stands casually in front of the castle, now watching intently, his mouth agape. While the first drawing was devoted to entertainment and sheer pleasure, the second shows everyday life: duck hunting, fishing, perhaps crab catching, not for sociable amusement but to feed the residents. People are doing everyday tasks. Despite the contemplative and peaceful atmosphere, it is an ordinary day, without the pronounced leisurely mood of the preceding scene. The entertainment consists of standing on the bridge and watching the activities.

The artist has closely linked the two contrasting drawings. The buildings could not be more different: here an airy structure with openings on all sides and figures obscuring the contrast between indoors and out; there the distinct solid cube of a water-surrounded castle whose windows do not reveal what is happening inside. The latter, on the other hand, has a wide open landscape. In the far distance is a castle on a hill and a town with a massive gate. Along the road towards it, a man crosses himself in front of a chapel. As for the bathhouse, the outside world is not wanted. A high wall strictly seals off the garden, and the only bit of the surroundings showing over the top are four identical trees. This ingenious contrast between the two drawings of course remains unnoticed by the casual observer. He is given a hint by the bather wrapped only in a towel whose glance leads from the first picture to the subsequent corresponding scene (fig. 27). She is differentiated from the other figures. As she steps into the bath, she seems to be thinking of something that is already on the next page. She mediates between the two pictures that are cleverly linked by their opposites – a device we will encounter again.

27 *A visitor to the baths,
detail from the bathhouse, fol. 19r*

Preceding facing pages:
26 *The water-surrounded castle,
fols 19v-20r*

Tournaments were one of the great forms of entertainment in the Middle Ages.[22] The *Manesse* manuscript gives a vivid impression of 14th-century ones (fig. 28).[23] In the late medieval period, jousting experienced a revival. Jousting societies were founded and tournaments took place on many occasions. During wartime or economic hardship, holding tournaments was not a high priority, as they were very expensive to organize and finance. Hence, after a series of great tournaments had taken place in the period before the Hussite Wars, it was only after a long pause that a major tournament was held almost annually from 1479 on.

Today's image of them is somewhat inaccurate, for the actual tournaments were a kind of mass brawl with clubs and dull swords. Two

28   *A medieval tournament, a miniature in 'Codex Manesse,' between 1300 and 1340, Heidelberg University Library*

Overleaf:
29   *The coronel joust, fols 20v-21r*

teams faced each other and tried, for instance, to knock the crests off each others' helmets.

Not everyone could participate. Naturally any kind of criminal offense – a marriage outside one's social class sufficed – disqualified the applicant. The basic prerequisite was aristocratic rank, which meant being able to furnish proof of a required number of aristocratic fore-fathers. A certain pragmatism cannot be denied: if the applicant married outside his class but into money – at least 4,000 florins – he could not participate immediately, but next time. Before a tournament a so-called display of helmets took place. Helmets and their crests were set up in a row and herolds verified whether their owners qualified to participate. If anyone did not, his helmet was knocked down to the ground. The kind of helmet used was called 'Spangenhelm.' It pro-tected the face with a metal grid, which gave the wearer a wide peri-pheral view to see his opponents. Indirect proof of one's qualification to joust consisted of evidence that an ancestor had participated in a tournament – thus one was qualified too. This was one of the reasons behind the rise of tournament books, which combined family history and tournaments. Altogether, participating in a tournament was an extremely prestigious affair, and it required considerable financial resources, down to organizing a suitable horse.

Besides the mass tournament, which was occasionally used illegally to settle old accounts, there was the 'Tjost,' a form of single combat. It was performed at weddings, for example, such as the famous Landshut wedding.

The Coronel Joust

The Housebook shows two tournaments. The first is known as 'Krön-leinstechen' or coronal joust, in which the aim is to knock the oppo-nent from his saddle by a powerful blow with the lance, without causing him serious injury (fig. 29). To this end, the thick lances have blunt ends topped by a small trident, the coronel. Moreover, the riders are armed to the point of immobility. On their heads is a heavy jousting helmet, or 'Stechhelm.' As its name suggests, it has a beak-like form offering minimal area for attack, but restricting vision consider-ably. It could only be used for this joust. The horse's forehead is also protected by armour, and a sack stuffed with straw is suspended in front of its breast.

The rider on the left has his lance in a couched position and is ready to gallop off, only restrained by the staff of a 'Griesswärtel,' a kind of steward who forms a part of the personnel at a tournament. The other knight, just being handed his lance, is urged to hurry. The device of letters and a crowned E on the shield and caparison has been interpreted as the name of the knight or, even less credibly, as the name of the artist.[24] Given the love of devices at the time, the straightforward label of a name seems too obvious. On the contrary, it was common to avoid any form of identification such as the family arms, even to ride under imaginary charges and mottos. For example, a stained glass window of a slightly later date shows a tournament with a knight whose helmet, shield and caparison are decorated with spectacles.[25] At the most, one used the initials of the beloved in whose honor one participated. It is probable that the initials here have some significance, it is just not clear what.

How difficult and often impossible deciphering such anagrams can be is demonstrated by the example of the device of Bernhard Rohrbach of Frankfurt (1446-1482): "Philip Katzman and Bernhard Rorbach had red clothes, breeches and balls covered with black fustian and lined in red taffeta, a light gray cape and on the right leg they had embroidered a silver scorpion and 4 silver Ms around it, and on the ball another silver scorpion and 4 vs on it, at the edge, and these mean 'Mich Mühet Mannich Male Vnglück Vntrew Vnd Vnfall'(I am oft troubled by sorrow, treachery and misfortune) and did it 'estomihi' 1472."[26]

The only combination of letters in the Housebook which can be allocated with certainty is the well known Habsburg A-E-I-O-U on a banner in the depiction of the army on the march.

The focus at the coronel joust is on the large audience. It includes children, young couples and a fool entertaining new arrivals with his flute. They are all watching and enjoying the performance.

The relationship between this drawing and another of the same period in the tournament book of Marx Walther has gone unnoticed

30 *The 1480 joust in the tournament book of Marx Walther, Munich, Bayerische Staatsbibliothek, Cgm 1930*

so far (fig. 30). He was born in 1456 into a wealthy Augsburg patrician family and was an enthusiastic participant in tournaments. His slim volume dating from the end of the 15th century contains pictures of twenty such events, though not all of them ended in his favor.[27] They show him making an impressive appearance in all different kinds of outfits, once with a little monkey, and once with a fourteen-year-old boy perched on the oversized lance to demonstrate that the monstrous weapon is not hollow. The book also contains a genealogy written by his father and copied in 1506, including a family tree as well as a list of the endowments granted by the Walther family in Donauwörth and Augsburg.

One of the tournaments, that of 1480, resembles the coronel joust in the Housebook in its layout, but its focus is on the participants. Not only the two men jousting – the opponent is, as is often the case, Jörg Hofmair of Augsburg – but also the stewards are all designated by name. They belong to respected Augsburg families, forming a merry throng in their red fools' costumes with long ears and sashes with bells. Two bagpipers and an umpire are also present. The event is structured in two centered rows. Musicians and the umpire form the top row, the contest begins below. The Housebook expands the contest by adding the audience in a third row.

The only steward between the armed knights in the Housebook, exhuberantly hurling his staff into the air, is in exactly the same position in the Walther book, here with a friend piggyback on his shoulders. The frontal figure of the umpire in the middle has its counterpart in the Housebook in the youth, somewhat off-center because of the fold, impudently grabbing his neighbor's breast.

Such a number of similar motifs is surely no coincidence. While the Housebook depicts many more figures and is far more skilled in handling them, both manuscripts are apparently based on the same model. As neither the Housebook nor the Walther codex can be dated precisely, the extent of their possible relationship to each other can no longer be determined. The artist of the Walther tournament shows closer affinities to a simple centered scheme, apart from the lively medley of red stewards. Stained glass roundels sometimes exhibit the same arrangement, with the umpire accompanied by musicians in the middle at the top (fig. 31).[28]

31   *Stained glass roundel with
imperial arms and joust, New York,
The Metropolitan Museum of Art,
The Cloisters Collection*

The Tilt

In the next tournament everything is completely different. Here the only spectators are warriors. There are no splendid caparisons, no ladies and no children in sight (fig. 32). Soldiers, for the most part faceless and anonymous, stand armed with crossbows as expert observers. A 'Scharfrennen' tilt is in progress and the opponents are both charging. They fight with sharp lances, wearing battle armour, i.e. gipons, or doublets that allow greater freedom of movement but more vulnerability. Swords hang at their sides as a matter of course. Umpires assign points according to the skill with which the opponent's shield is hit. This kind of joust, shown here in its basic form, could easily be organized ad hoc in an army camp. It was a favorite exercise for young hotheads who could not afford their own jousting gear. A horse race sweeps across the top of the page, accompanied by a barking dog and an unwelcome participant.

Less conspicuously, not as obviously as in the first pair of drawings in this series, the Housebook Master has cross-referenced the compositions of these two scenes as well. He does so by mirroring the two tournaments horizontally. While musicians and a few mounted spectators at the coronel joust mill about in the top row, at the tilt they have their backs turned and stand in the foreground, this time in the bottom row.

Overleaf:
32   *The tilt, fols 21v-22r*

The Medieval Housebook

Facsimile and Commentary

Prestel

This famous work of medieval drawing offers a unique and vivid insight into the way of life and thinking in the world 500 years ago

Incomparable in its vivid portrayal, the famous 'Medieval Housebook' provides a humorous insight into the world of the late Middle Ages in Germany. In addition to a discourse on mnemonics, a section on astrology with drawings of the planets and on the life of the knights — complete with tournaments and gardens of love — medicinal and household recipes can also be found as well

The collection

The foundation of this important collection belonging to the Princes of Waldburg Wolfegg was laid in the mid-17th century by the Lord High Steward Maximilian Willibald (1604-1667). In contrast to the majority of important royal collections of engravings and drawings, the Wolfegg cabinet did not pass into a public museum but has survived to this day in its traditional home in the print room at Schloss Wolfegg.

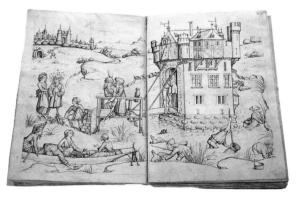

In front of a castle surrounded by water — day-to-day life goes on, with duck hunting and fishing (fol. 19v-20r)

Preparations for 'crown stabbing', a tournament in which the opponents try to knock one another from their horses with lances (fol. 20v-21r)

as chapters on mining and the art of warfare. Thus the manuscript sketches a wide-ranging picture of thought and knowledge at the time. The Housebook, completed in 1480, is particularly important due to its artist, who portrayed his world with wit and sensitivity. The so-called Housebook Master,

Mining panorama

who worked in the Middle Rhine region and is known from several other works, had considerable influence, particularly through his engraving, on his younger contemporary, Albrecht Dürer.

The identity of the artist as well as the patron remain one of the greatest mysteries of art history.

The contents of the 'Medieval Housebook'

Artists
The 'artes liberales':
  Mnemonics
  Drawings of planets
The life of the knights:
  Bathhouse and
   moated castle
  Tournaments
  Hunting and love
  The obscene garden
   of love
Household remedies
Mining, smelting and
  minting
The art of warfare

Troubadours and artists

For over 300 years, the manuscript has formed part of the collection of the Princes of Waldburg Wolfegg at Schloss Wolfegg near Ravensburg.

The original of the 'Medieval Housebook' will be on display in the following locations:

Städelsches Kunstinstitut, Frankfurt a. M. (18.9.-2.11.1997)
Rijksmuseum Amsterdam (22.11.1997-18.1.1998)
Haus der Kunst, Munich (24.7.-11.10.1998)
National Gallery, Washington (15.11.1998-31.1.1999)
The Frick Collection, New York (11.5.-25.7.1999)

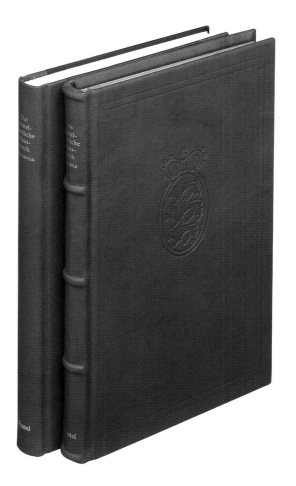

*The 'Medieval Housebook'
as a facsimile with a commentary
volume in a strictly limited edition*

The Medieval Housebook
Das Mittelalterliche Hausbuch

Limited facsimile edition with commentary volume
in 750 numbered copies.

Edited by Christoph Count of Waldburg Wolfegg.
With contributions by Gundolf Keil, Eberhard König, Rainer Leng,
Karl-Heinz Ludwig and Christoph Count of Waldburg Wolfegg.
Format 7¾ x 11¾ in. ISBN 3-7913-1838-1.
The volumes are only available together in an ornamental slipcase.

To give an impression of the high quality of the edition, a file
is available containing three pages reproduced in facsimile in
the original format, together with a detailed description of
the facsimile for a token fee of US $ 65.00. When buying the
facsimile edition, the price of this file will be deducted.

The edition:
The two-volume publication will appear as a
facsimile with a commentary volume in a strictly
limited edition. Volumes 1–750 will be numbered
by hand in Arabic figures and intended for sale.
In addition to these, there will be 50 volumes
numbered in Roman numerals which will not
be available on the market. When printing is
complete, the printing plates will be destroyed,
thus guaranteeing the uniqueness of the edition.

Facsimile volume: To reproduce the facsimile, the original
manuscript from Schloss Wolfegg has been taken apart by experts. The 130 pages have been reproduced with the utmost
care and printed in five colors (with gold). The printed sheets
are cut individually and bound by hand. The full leather
binding is embossed with gold on the spine and, on the front
cover, with the coat of arms of the Princes of Waldburg.

Commentary volume: The accompanying book is published
bilingually in German and English. Approx. 236 pages and
48 b/w illustrations. Half-leather binding with embossed
gold on the spine.

The first colored 'Housebook' facsimile: a must for all collectors and lovers of book illumination

◄ In the 'Medieval Housebook', seven planets appear at the beginning, arranged according to their periods of revolution. These affect the character of the individual at the hour he is born. The last planet with the fastest period of revolution is the moon. The children of Lady Luna are thus predestined for careers linked with water, such as fishing or milling, or something suiting their changeable and freedom-loving natures, e.g. bird-catchers, magicians or artists.

"The stars have worked their magic on me, I am unsettled and strange. My child can barely be tamed, she will be subject to no one… And he who makes his living from water, shall be blessed by the light of the moon."

Following the tour of international exhibitions from 1997 to 1999, the priceless original of the 'Medieval Housebook' will be reassembled and newly bound, to be locked in the safe at Schloss Wolfegg.

The famous manuscript will subsequently only be accessible in the form of this first colored facsimile. The edition will thus become a very sought-after collector's piece and an invaluable source for research.

Prestel   Mandlstraße 26   80802 Munich   Phone 089/3817090   Fax 089/335175

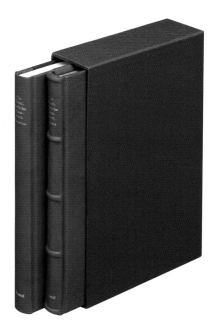

Order Form

ISBN 3-7913-9967-9, 5 Pera 798

The Noble Hunt

The tournaments are followed by the third and last pair of parallel pictures. Their subject matter comprises two different kinds of hunting. On the first double page we see a large-scale landscape and in the background a castle, a village and the bent figure of a peasant who, having broken out in a sweat from plowing, is pulling his smock off over his head (fig. 33). In the foreground there is a merry carefree hunt on the noblest species of game: the stag. While the does have sought shelter in a wood, the stag collapses before the company of hunters. The leader of the company, riding without an accompanying lady, points in the direction of the game. He bears the device of the Order of the Jug. This hunt is a relaxed affair. No one carries weapons, at most a whip to guide the horse and the pack of hounds. The ladies sit behind the lords on the same horse, fearing no danger. There has thus been no riding up hill and down dale, and little risk of falling from one's mount. Their attentiveness leaves something to be desired: one of the participants points out to his companions the comical figure cut by the peasant with his contortions in the field. This is another example of the device that the Housebook Master contrived in order to link groups of figures thematically. Thus a youth in the first tournament was less interested in the incipient joust than in the bosom of a girl he was brazenly groping, despite the presence of his lady behind him on his horse. In the scene to come of the army on the march, curious soldiers turn to stare at a drunkard threatened by his wife about to take away his bottle (fig. 72).

The refined form of the mounted hunt is totally abandoned in the next pair of facing pages (fig. 34). This chase takes place around a castle with many buildings. The wall is no defensive rampart but makes a welcoming impression. The open stables and the activities taking place in and around them invite us to approach. The women at home here get right down to business: they are hunting men. A buxom woman behind the well beckons to a stableboy. Another has tucked up her skirts to be able to run faster. Appropriately enough, she carries an empty cage in her hand. A more proper lady places a cautionary hand on her arm in a fruitless attempt to restrain her. The object of her desire is a peasant dangling helplessly in the air from a trap. The peasant's agitated wife gesticulates in a field in the background (fig. 35).

Two couples are already intimately occupied, one wandering along with their arms around each other, the other in a close embrace. The drawing is covered with allusions, such as the stork clattering from the rooftop in mating season, and the decoy attracting entire flocks of birds. A similar scene with a fowler appeared on the Luna page. The older ladies in this disreputable world either hinder the efforts of the younger women or are left to do housework on their knees.

As opposed to the stag hunt, this chase is dangerous for the men. The stag is noble game, which refers less to its size – the wood grouse and eagle also qualify as noble – than to the social rank of its hunters. The stag hunt was a high, aristocratic hunt, long restricted to the classes in power. Only hunting with a falcon had equal status. There are several instances of the bird accompanying young nobles in the Housebook.

How to solve the riddle posed by this pair of pictures is indicated by an allegory of courtly love, *Die Jagd (The Hunt)*. It is the only known work by Hadamar von Laber (*fl. c.* 1300 – after 1354), scion of aristocratic stock from the Upper Palatinate. The poem, dating from the second quarter of the 14th century, survives in eighteen manuscripts, indicating that it was a work of some importance in the late Middle Ages. Laber was highly esteemed both as a poet and an expert authority.[29]

The allegory describes how a mounted huntsman sets out in pursuit of noble game with his hounds, named "Joy", "Will," "Delight" and "Loyalty," with the lead dog, "Heart." They succeed in finding the doe, but the lead dog is wounded. At the second try, after an attack by wolves, the dogs "Kiss" and "Embrace" are let loose. But the huntsman refrains from employing the dog "End," with the result that the doe escapes. "Heart" is again sorely wounded. The final 200 of the total of 565 stanzas tell of the lover's sorrow: his adored lady withdraws more and more, her trail is lost, yet he remains loyal forever.

Overleaf:
33   *The noble hunt, fols 22v-23r*

Preceding facing pages:
34   *In pursuit of lesser game, fols 23v-24r*

35   *The trap, detail from*
'The Pursuit of Lesser Game,' fol. 23v

Correct behaviour is repeatedly addressed by means of contrasts. For the hunter this means that he should be committed only to noble game, and that it is inadmissible to abandon the trail in favor of another. Elsewhere, two huntsmen discuss which forms of the hunt are permitted. Only hunting according to courtly rules is allowed, that is, without tricks or violence or even traps and nets, in order to bag, as it were, the aim of one's desires.

The hunt and hunting motifs such as the find, the watch, the necessary endurance and patience as symbols for the pursuit of love occur frequently in the visual arts, for instance on jewelry boxes and the backs of mirrors. Playing cards used in aristocratic circles also revolve around the theme of the hunt. A luxurious deck of cards preserved in Stuttgart, designed *c.* 1430 in the Meuse area or Upper Rhineland, has the hound, falcon, stag and duck as its suits. The theme of love is suggested by the elegant 'queen' playing suggestively with a stag, as if it were the unicorn that can only be captured by a virgin (fig. 36).[30]

The drawing of the chase in the Housebook goes beyond the poem by Hadamar von Laber in that it deals not with men resorting to devious means but with women. Furthermore, their enticements amount to a pursuit of lesser game. They exhibit a vulgar attitude and lust. Nor are their methods very dignified. Deception, tricks and ambush are the order of the day. This hunt does not serve to entertain but to satisfy earthly desires. The sprung trap, the empty birdcage and the decoy indicate that here the men are lesser game, no better than the birds or rabbits that fall into a trap.

Hadamar von Laber's hunt is not as rich in imagery as this short summary might suggest. He does not describe the green of the forest,

36   *Lady with a stag,*
a playing card from the Stuttgart Deck,
c. 1430, Stuttgart, Württembergisches
Landesmuseum

the appearance of the people or the animals, and he only alludes indirectly to the surroundings and events through the reflections of the huntsman and the changing composition of the pack of hounds. Laber does not conform to established rules or a clearly defined image such as in the *Roman de la Rose*. The only fixed point is the steadfast loyalty of the hunter, which prevails even in the face of the complete lack of any prospect of success. Püterich von Reichertshausen, a man of considerable education and our source of information about the author of *The Hunt*, regretted that Laber was no longer alive in 1462. He would have liked the author to provide "a gloss upon his noble poem."[31] Besides expressing his admiration for the refinement of the poem, he thus admits that he is unable to understand the allegory.

We find ourselves in the same situation as Reichertshausen. A gloss on the Housebook would be most welcome. Even the interpretation of individual pictorial elements does not clearly explain the whole. An owl, for example, can have a completely different meaning from one picture to the next. Some day an allegory of courtly love might be found that exhibits even closer correspondences to the Housebook series. However, I doubt that a literary source was directly transposed into a visual image. Features common to the Housebook's drawings and the literary works need not have resulted from direct influence. They might be explained by the fact that both genres create imaginative variations on an old theme while breaking new ground.

The correlation of these pairs of courtly scenes becomes increasingly ingenious as the series proceeds. While the first pair contrasted no more than open or closed buildings and corresponding landscapes, the contrast between the tournaments in the second pair was heightened to a mirror effect. The final pair is skilfully linked according to a cleverly contrived scheme in that the artist again mirrored the composition, this time along the vertical of the fold instead of the horizontal. Couples who have already come together sit on horseback on the right in the first scene, and wander on the left in the second. The rider pointing to the stag has his counterpart in the woman beckoning to the 'game,' i.e. the stableboy, in the second picture. The landscape opens out far into the distance in both, here on the right, there on the left, where a grotesquely contorted figure appears in both as well.

Its correspondences in content and form bestow an artificial character on 'The Pursuit of Lesser Game.' Unlike all the other drawings in the Housebook, it is not a descriptive and apparently realistic representation that makes a harmonious impression in itself even without our knowledge of its intellectual background. That is why it was always considered a riddle, remaining unsolved to this day.

Looking at this series of six scenes as a whole, we discover their meaning from their arrangement in pairs. Their motto could be 'dream and reality.' Both dream and reality are reflected in the longing for a garden of delights, the desire for knightly games and the reality of war, the aspiration to courtly love and the ignoble hunt.

The Obscene Garden of Love

While the previous six scenes turned out to be three pairs of pictures that belong together, the final drawing in this gathering stands alone. As does the first scene of this sequence, it plays in an enclosed garden (fig. 38). In this garden of love, however, dining is an occasion for vice and people are just plain lecherous – a logical progression from the preceding 'Pursuit of Lesser Game.' A motley company is displayed like colorful beads on a glowing yellow ochre meadow. A brilliantly hued peacock perches on a fence. On the right a bubbling stream of ice-blue water limits the garden. Only a little bridge made of two thick planks crosses to the far side. There, steep cliffs tower beyond a pumping machine driven by a water wheel, of the type that serves to pump out water accumulating in a mine.

Here again the Housebook Master has undisguisedly drawn on an engraving by Master E. S.: *The Large Garden of Love*, where the dining company is closely framed by a fence (fig. 37). The man entering, complete with gate and the entire company have been borrowed almost down to the last detail. Everything else is new, including the woman in the foreground, pointing to the exposed genitals of the fool, and her buffoon of a companion. Minor changes were made: the plate holds more pears than apples, and the headgear of some of the figures was elaborated so as to fill the space created by moving back the fence.

The effortless ease with which a new couple was added to the group reveals the expertise of the artist. He adapted the whole panorama to the scene he took over from Master E. S. He created a counterpart to the engraving's high-contrast pattern by darkly shading the cliff and he distributed fluffy trees and bushes throughout in order to keep the composition balanced as a whole. The coloring has been completed with body color. The drawing gives the impression of an alien element compared to the others in this manuscript. It is conceivable that an earlier source is likewise behind the other folios in the Housebook that are often attributed to other artists. This would explain the incongruent impression made, for instance, by the mining panorama with its strangely round-headed figures (fig. 44).

A fountain forms the finale of this series of pictures just as the fountain in the very first scene, admired by the ladies entering the garden, formed the prelude. The rushing brook represents a break and a division. It can only be crossed via the narrow bridge leading to the other side of the river, to new subject matter. The waterworks there, in

Overleaf:
38 *The obscene garden of love,*
fols 24v-25r

37　'The Large Garden of Love,' engraving by Master E.S.

the form of a pumping machine used to drain groundwater from mines, points to the next series of illustrations, that of technical devices.

The Knight of the Order of the Jug

Remarkably enough, religious life is only referred to incidentally in the Housebook. In the background of the picture of the water-surrounded castle a wanderer crosses himself in front of a chapel, and among the children of Sol are worshippers at an altar in church. Not one chapter, not even a single page, is devoted to the faith despite the fact that religion controlled much of life at the time. There is, however, an indirect reference to religious life. One of the noblemen wears a narow white band over his left shoulder which is embroidered with the emblem of a jug. He appears on almost every page, usually in a prominent position (see fig. 39). The emblem is the device of the Order of the Jug mentioned above, which was founded in the 14th century by Ferdinand I of Castile and Aragon (1379/80-1416) and revived by Frederick III in about 1473.[32]

In the course of the 14th century and becoming increasingly widespread in the 15th, a new form of the Order developed that led to the mere wearing of the badge. The vows were abandoned in favor of a more loosely connected society. Large numbers of members were eligible and concurrent membership in other related orders was permitted. Such societies were primarily committed to defending the true faith, as was the Order of the Dragon founded by Sigismund against the Hussites, as well as to helping the needy.

The statutes of the Order of the Jug date from before 1458 and have survived in a copy. The second regulation reads:

"They are to wear the device each and every Saturday, but if necessity prevents them from wearing the whole device then they are to wear a part; moreover, everyone may choose whether to be dressed entirely in white every Saturday, or to wear a stole or ribbon, three fingers in breadth, unadorned in the middle save for white pearls or something else white in color. At the hems and around them any color may be worn, provided in the middle of the stole there remains an area of white three fingers in breadth, and they are also obliged to wear the device on all holy days devoted to Our Lady, as on Saturdays."[33]

Thus the male members of the Order were required, on Saturdays and feast days dedicated to the Virgin, to dress entirely in white or to wear a white stole or band, three fingers wide. This was decorated with the device, or symbol of the Order, a jug-shaped vase with three lilies. The vases could also form the links of a chain from which hung a griffon, whose white wings could be gilded as a reward for some deeds

39 *The Knight of the Order of the Jug, detail from the bathhouse, fol. 18v*

performed "against the unbelievers." According to the statutes, the device was to be worn all one's life. The origin of the symbol referred to the obligation to be devoted particularly to the worship of the Virgin Mary. The original Spanish motto was: "por su amor," "for love of her," the mother of God. The annoying word "jug" in the name of the Order refers to the vase with lilies traditionally displayed in representations of the Annunciation, which are a symbol of the purity of the Virgin. Under Emperor Maximilian I the motto was later changed to "Halt Mass" (observe moderation). Dürer's woodcut for his *Triumphal Arch* shows a chain of the Order of Moderation with a medallion depicting the Madonna and Child in a mandorla (fig. 40). An aspect of the lily which clarifies the later motto is the reference to moderation as "harmony in both the cosmic order and human behavior."[34]

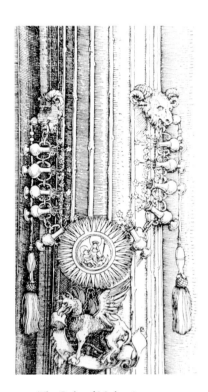

40   *The Order of Moderation, detail from the 'Triumphal Arch of Emperor Maximilian I,' woodcut by Albrecht Dürer, 1515*

The Order of the Jug was granted to a many people from both the upper and lower aristocracy and to patricians. A 1459 document of the conferment of the order upon Nicholas and Sophia von Lobkowicz by Frederick III has survived. The minnesinger Oswald von Wolkenstein was also bearer of the Order of the Jug, which he proudly wears in two portraits. The Augsburg patrician Sebastian Ilsung was accorded the order during a pilgrimage in Spain.[35] Different versions of the device were in circulation, as it was granted both in Spain and in the German-speaking countries. In the context of our manuscript there is no connection with Spain. Hence this order must have been granted by Frederick III, who wore the Order of the Jug himself at his meeting with Charles the Bold in Trier in 1473.[36]

The repeated appearance of the motif in the Housebook is in any case a clear indication of the pride felt by this bearer of the order. The Knight of the Jug is often dressed in red and green, a color combination that is common in the Housebook. While unable to make a proposal of my own, I doubt whether the combinations of red, white and green or red and green do in fact indicate the colors of Augsburg, as Retberg as well as Bossert and Storck suspected. Their idea stems from a time when the Housebook Master was still considered to be of South German origin.

Although a series of drawings are bracketed together by the recurring appearance of the bearer of the Order of the Jug, they do not represent various episodes of a narrative. The Knight of the Jug may have served to emphasize a courtly element. It is more likely that his presence is a personal reference to the patron, who was conceivably a member of the order himself.

The Knight of the Jug can be traced by means of his device in only the positive sections of these pairs of pictures. First he escorts a couple into the garden (fig. 25). Then, with a very similar physiognomy, he is the second of the knight with the arms of a crowned E (fig. 29), and finally, he is the leader of the noble hunt (fig. 33). In the other scenes he can be surmised twice, but it is only his scalloped sash that shows. In 'The Pursuit of Lesser Game' he has his back to us (fig. 34) and in 'The Obscene Garden of Love' his arm conceals the device (fig. 38). In the same picture, clearly identifiable once more, the Knight of the Jug walks with a lady who points to the lascivious company, safe on the far side of the brook.

The next section, taking up sixteen pages (fols 26r–33v), contains innumerable recipes, some of their decisive ingredients encoded in Hebrew letters. They are not in a random accumulation but thematically arranged. This is rare in such collections of lay medical texts, which were very popular at the time and normally merely listed the prescriptions in no order whatsoever.

41　*The spinning wheel, fol. 34r*

Wiltu guten essich machen

Nym wein ber vor sant michels tag thu die in ein verglast gros geschirre setz an die sonne las in vergeren dann thu die trüsen dar aus vnd geuß das vff in ein clein fesslein das si geschmecket sey mit guten wein vnd dü den vergerten wein dar in Also das es 2 finger dieff wan sey vnd verschwint es wol vnd lass es sten an der worm einen monet so hastu de aller storckesten besten essich

Biber schwartz machen.

Beytz den biber sie in dampf zwikapfel mit eyn wenig dann öles biß sie weich fallen Nym zu 1 lib d öpfel 2 lot alun 2 lot es stu 1 clein salor : machs als du weyst

Aus flachs side

Mach ein starck laugen von wand aschem vnd calvi saltz siede mit ein vnd kluß die laugen durch ein filtz Nym dann gehechelte flas loß in sieden dar inn 2 stunt wort das mit uber gee wasch in dann hut er mit gemach siede in mee

Duch linen willen vnd flachs.

Swartz verben zu einer ein gehort 2 lot ... dü sie in eym siedemges wasser vnd 1 lot alun dar zu vnd darnach das duch loß sieden 1 stunt vnd reuher vnd kile es dann ab Nym dann 3 lott vitriol 1 lot grüme fingel stein von noch gedrincten loß sieden vnd kil es d weil sieds biß schwartz gemig wirt

Ein wasser zu flecken

Zu den gemende wie ma die aus brennzeit blym zu greyndern stall nach einst so vn gemeynes wasser 1 lib wein stein 2 lot alun stoß clein loß sieden 1 lym danne ein trinck glass vol essig geuß thu dar inn 1 lot mith geuß dar zu loß das dritteil in sieden

Rot ferben.

En geyssen stocken schneyt sie clein vnd seude sey in reimwasser stoß sie vnd das sie zu klumpen wirt off ein roiß sie wol drucken So du wdarefft B stoß sie clein vor mach den morsel das das best mit verstede dann mach ein wasser als wym wassen kwen seut wasser dar mit bisses auff seudt Schuttes dan in ein ede treschnere vnd ceue es also wol durch einander biß es plume rest dann deit es zu es müß zum wynsten ston e tag gefallen dann nym das o berst wasser Thu es in einen kessel wann es auff seudt So thu der vorderen ferben dar inn zu dryn clein 1 lib vnd dan das drin loß siedem 1 fiertel eine stunde vnd dan drucken vnd wider in die selb farbe diß es biß es schon rott wirt vnd wasch es Das duch das du rot wolt ferben Sib in var also bereiten Nym alun vermasch loß es herwallen thoß das duch oder seyden darin vnd laiß wie clem noch dar inn sieden 1 lym dann rocr in stem stoß in auch clem sende in auch mit regen wasser zu 3 eln 1 fiertel stoß das duch dar inn vnd wol auß vnd loß drucken farbe es dann in der vordern farbo so ist es bereit

The treatment of wounds is first, along with instructions on everything from treating fractures to stanching bleeding. Next, internal medicine covers several pages, including some prescriptions for laxatives as well as cures for diarrhea, the treatment of hemorrhoids, remedies against plague and many more. Then there are two recipes for perfumed rosary beads and one for confectionery that induces women to sleep with men, its main ingredients being nutmeg oil and eggyolk (fol. 31v). Methods for hardening iron, dyeing fabric, and measures against bedbugs and moths can also be found. Finally, there are two recipes for cooking: a baked dish and a cheese and egg pancake.

The home remedies are followed by a single picture from the domestic domain: an unusual full-page drawing of a spinning wheel. It is now thought to be the earliest representation of the revolutionary apparatus that simplified work considerably (fig. 41). Since time immemorial, people had worked with a simple spindle turning on the floor to wind up the yarn. The handwheel came to facilitate the procedure, with the disadvantage that work had to be interrupted regularly because the wheel could only either spin or reel. In the Housebook the mechanism is equipped with a U-shaped hook with eyes that winds the yarn onto a spool during the spinning process. The only thing missing is the treadle drive. What is significant is that a spinning-wheel is depicted here in the first place. The patron was surely not someone who would sit at one himself. It was the technology that fascinated him. The ingenious mechanical solution and the state-of-the-art technology mattered more to him than any practical applications for himself.

A different tone prevails before the gap of the two missing gatherings from that in the section after it. Was the first half of the book originally also divided into two chapters covering three gatherings each, corresponding to the second half dealing with mining and war? Whatever the case may have been, the correlation between the texts and images after the big gap becomes unclear. The home remedies and the spinning wheel have practical matters in common. How they could be reconciled with the highly original depictions of imaginary scenes of courtly love and life in the days of chivalry must remain an open question.

42   *A page with recipes for dyeing cloth and removing stains, fol. 32v*

MINING,
SMELTING AND MINTING

In its reconstructed state a good half of the Housebook covers techno-
logical subject matter. Technical sobriety comes as a surprise after
the lively scenes of merry activity. It is introduced by a mining panorama facing the second heraldic page. Bright and cheerful, it is
imbued with soft greens, the mountains fading into the hazy blue
distance (fig. 44). The composition revolves around a central conical
rock. Approaching from a valley on the right, where a town lies on a
river in the distance, are several riders and a loaded cart. Miners work
in the dark tunnels in the rock, others collect and sort the yield. On the
left, wheelbarrows are pushed along a path that continues up to a castle
on the horizon.

This folio illustrates different aspects of mining, from the hauling
and crushing of ore to an imposing administration building, which
may also contain an assaying furnace. Four ruffians with daggers and
swords are engaged in a murderous fight in front of it. Adjacent but
separated by a strip of lawn, stands an elegant couple. The gentleman
wears a cap with a large tassle and carries a thin cane recalling a
teacher's pointer. He turns towards a lady and is apparently explaining
the work to her. The two of them recall the spectators around the
bearded master in the drawing of the artistes introducing the first part
of the Housebook. They do not participate either, but are impartial
observers. Once again, the man is a Knight of the Order of the Jug,
encountered so often in the scenes of courtly love and chivalry. While
there he accompanied a lady into the garden of love, here he intro-

45   *Interior of a smelting plant with
cupel furnace, bellows and workers,
fol. 35v*

Preceding facing pages:
43   *The coat of arms of the patron,
fol. 34v*
44   *The mining panorama, fol. 35r*

duces her to the activities in a mining landscape. One page further on we will meet him again with his companion in a smelter.

On what originally comprised three gatherings, there now follow drawings and texts on mining and mintage. The drawings are among the earliest representations of the subject. They deal with silver mining, the most profitable of businesses. Before smelting could take place, the ore had to be crushed in order to separate the minerals from the barren rock. The ore was then pre-sorted according to its degree of purity. The more the usable and useless elements were intermingled, the longer it had to be crushed in order to separate the two components. This was accomplished by stamping machines such as the one shown on fol. 36v (fig. 51).

During the next stage the different constituents of the mineral ore were separated chemically. On fols 35v and 36r, facing illustrations show the interiors of two different smelting works where the ore was refined in furnaces. The furrow drain process shown here was a technique whereby silver could be obtained from argentiferous copper ore by adding lead.[37] This was doubly profitable, yielding both pure copper and silver. The first interior shows a so-called ore calcination furnace on the left which was alternatingly stocked with ore and coal (fig. 45). This first basic refinery process served to remove unwanted sulfur and arsenic. To the right of it is a round cupel furnace, in which lead was drained off the molten silver. Two bellows for the fire are operated from behind a protective wall against the heat. The people are entertainingly portrayed, both those who work here and those who visit. They are not mere impartial props: by holding his cloak in front of his face, one of the men tries to protect himself against the heat and the noxious fumes emanating from the furnaces.

46   *Interior of a smelting plant for the furrow drain process and tools, fol. 36r*

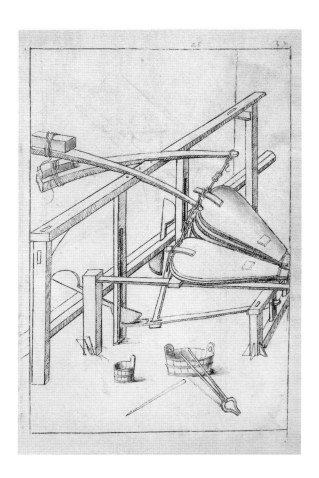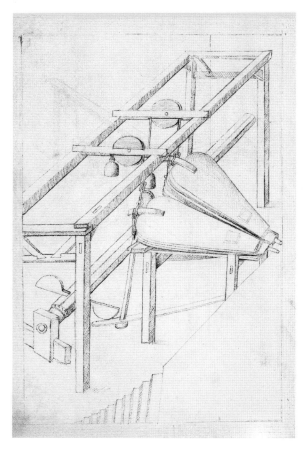

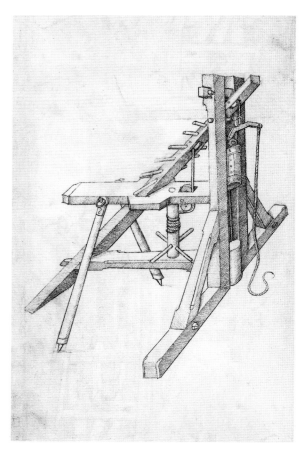

The drawing opposite is like something from an instruction manual, showing tools spread out on the floor (fig. 66). This apt assemblage of basically boring equipment into an attractive composition is outstanding in quality for the period of the Housebook. It sets itself off from the monotonous or clumsy rows of tools in other manuscripts that are scarcely more than catalogs. The round furnace is presumably again for cupellation. Two shaft furnaces tall as a man have open 'eyes,' or tapping apertures, through which the metal was directed into crucibles, from whence it poured through a drain into assembled pans. The furrow drain furnace on the far right served to extract argentiferous lead from copper ore.

The furrow drain process is considered a Nuremberg invention, perfected by about 1450. The smelting plants in and around Nuremberg were soon moved by order of the town council to the Thuringian Forest. Environmental pollution was less of a nuisance there, and plenty of wood fuel was at hand. But the artist could have found such plants in the lower Inn valley as well – should he have been familiar with them from personal experience.

Smelting Recipes

In the smelting recipes that follow, the names of the metals are replaced by the corresponding alchemistic signs (fols 40r–41r). The alchemistic origin of many of the methods is worthy of note, such as assaying ore with antimony glance or refining gold by using antimony. One section discusses how the color of the flame can serve to identify the various metals in a smelt. Another below deals with the extraction of alum from alunite, which came from central Italy and was not mined farther north until the time of the Housebook.

A short text on minting follows, aimed personally at someone wanting to start a mint and therefore needing to know about wages, materials, and taxes and duties. In the second half of the 15th century the market was in a state of disorder due to increased silver mining and the proliferation of inferior coins. A decree by Emperor Frederick III of 1442 prescribed 19-carat coins as the norm. (A carat is a measure of the proportion of gold in an alloy. Thus 18-carat means that 750 of a total of 1,000 parts consist of fine gold.) Nevertheless, 19-carat gold currency existed only in name by the end of the century, for a gold coin was now more expensive than its regulation equivalent value in silver coins. 'Commercial florins' made of silver instead of gold now became common. The tables in the Housebook range from 12-carat to pure 24-carat gold (fig. 52). To use them, the minter first needs to determine the fineness of the gold coin, done most easily with the following test: A line is drawn by running the coin across a black sam-

Top left:
47   *Twin bellows with counterweights attached, fol. 37r*

Top right:
48   *Twin bellows with metal weights running on rollers, fol. 37v*

Lower left:
49   *Encased stamping machine or pumping plant, fol. 38r*

Lower right:
50   *Vertically adjustable pile driver, fol. 38v*

pling stone. Acids of different strengths are applied and, depending on the gold's fineness, the line disappears or remains. Once the fineness and the weight of the pure metal are determined, the buying rate is found with the help of a table. Hence, for example, gold with a fineness of 22 carats and a weight of 4 lots costs 19 florins and 5 shillings. Unfortunately, careless mistakes increase as the table for silver proceeds and the list eventually breaks off, no doubt because the scribe realized that he had overshot his mark and his table could be continued infinitely.

One special feature is not obvious to us today because it has become taken for granted: computing in Arabic numerals. This was just beginning to become common in trade and in monetary transactions at the time. In all other domains arithmetic continued to be done in Roman numerals. It was not until the book by Johannes Widmann, *Behende und hübsche Rechnung auf allen kauffmanschafft (Useful and Elegant Arithmetic for all Commerce)* was printed in Leipzig in 1489 that the simpler way of calculating became widespread. The compiler of the Housebook's tables of weights and of most of the other texts was therefore up-to-date when it came to modern business practices.

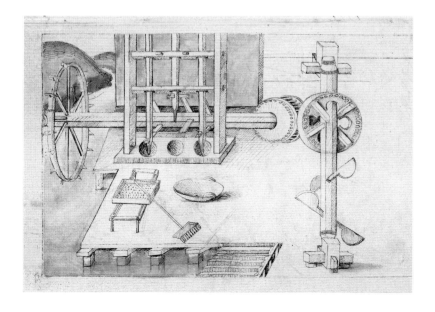

51 *Stamping machine driven by a water wheel with equipment for rinsing ore and a transmission, fol. 36v*

Opposite:
52 *Tables of weights for gold, fol. 43r*

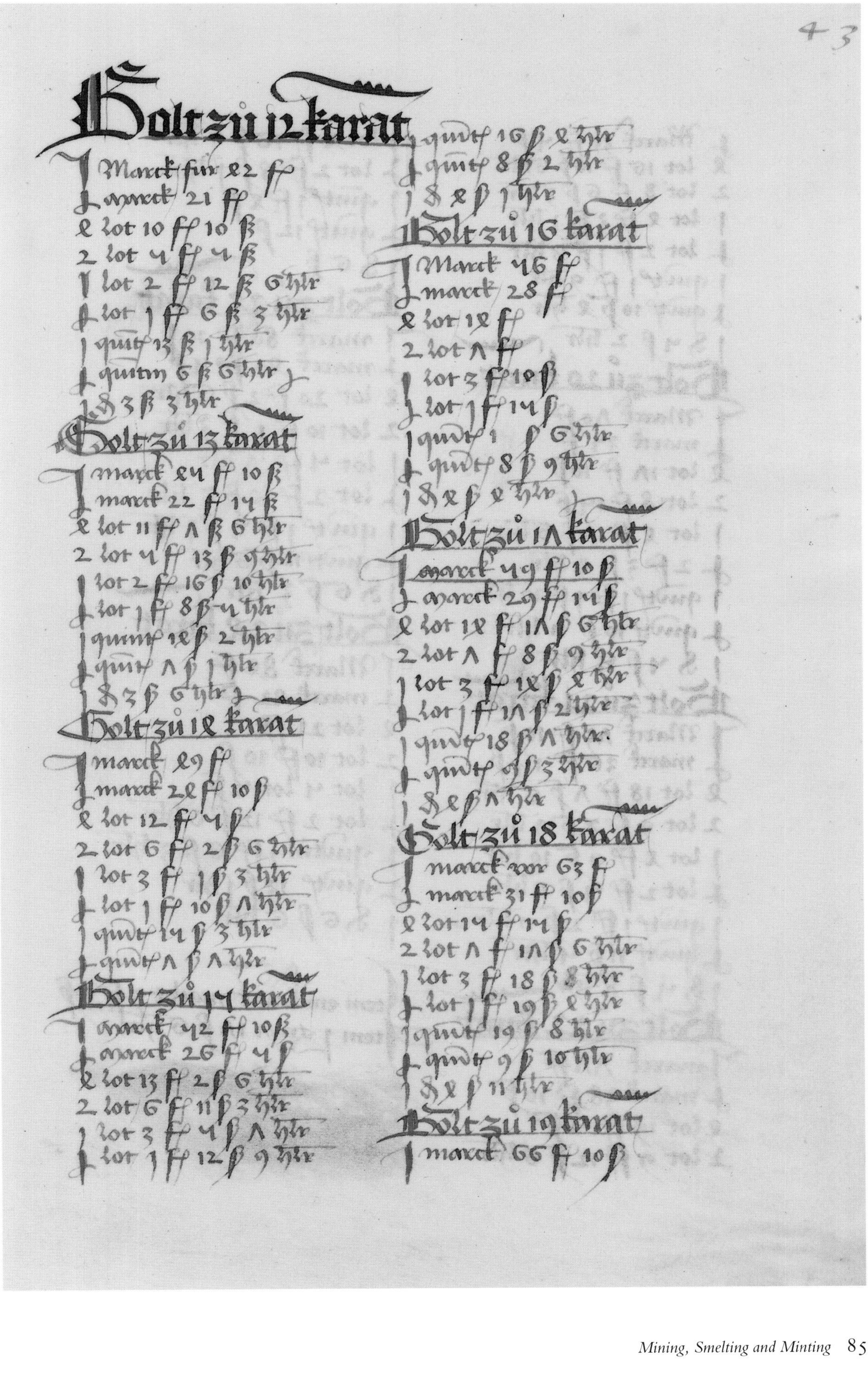
43

Holt zu 12 karat
1 Marck fur 42 fl
½ marck 21 fl
4 lot 10 fl 10 ß
2 lot 5 fl 5 ß
1 lot 2 fl 12 ß 6 hlr
½ lot 1 fl 6 ß 3 hlr
1 quint 13 ß 1 hlr
½ quintin 6 ß 6 hlr
1 d 3 ß 3 hlr

Holt zu 13 karat
1 marck 45 fl 10 ß
½ marck 22 fl 15 ß
4 lot 11 fl 7 ß 6 hlr
2 lot 5 fl 13 ß 9 hlr
1 lot 2 fl 16 ß 10 hlr
½ lot 1 fl 8 ß 5 hlr
1 quint 14 ß 2 hlr
½ quint 7 ß 1 hlr
1 d 3 ß 6 hlr

Holt zu 14 karat
1 marck 49 fl
½ marck 24 fl 10 ß
4 lot 12 fl 5 ß
2 lot 6 fl 2 ß 6 hlr
1 lot 3 fl 1 ß 3 hlr
½ lot 1 fl 10 ß 7 hlr
1 quint 15 ß 3 hlr
½ quint 7 ß 7 hlr

Holt zu 15 karat
1 marck 52 fl 10 ß
½ marck 26 fl 5 ß
4 lot 13 fl 2 ß 6 hlr
2 lot 6 fl 11 ß 3 hlr
1 lot 3 fl 5 ß 7 hlr
½ lot 1 fl 12 ß 9 hlr
1 quint 16 ß 4 hlr
½ quint 8 ß 2 hlr
1 d 4 ß 1 hlr

Holt zu 16 karat
1 Marck 56 fl
½ marck 28 fl
4 lot 14 fl
2 lot 7 fl
1 lot 3 fl 10 ß
½ lot 1 fl 15 ß
1 quint 1 ß 6 hlr
½ quint 8 ß 9 hlr
1 d 4 ß 4 hlr

Holt zu 17 karat
1 marck 59 fl 10 ß
½ marck 29 fl 15 ß
4 lot 14 fl 17 ß 6 hlr
2 lot 7 fl 8 ß 9 hlr
1 lot 3 fl 14 ß 4 hlr
½ lot 1 fl 17 ß 2 hlr
1 quint 18 ß 7 hlr
½ quint 9 ß 3 hlr
1 d 4 ß 7 hlr

Holt zu 18 karat
1 marck vor 63 fl
½ marck 31 fl 10 ß
4 lot 15 fl 15 ß
2 lot 7 fl 17 ß 6 hlr
1 lot 3 fl 18 ß 8 hlr
½ lot 1 fl 19 ß 4 hlr
1 quint 19 ß 8 hlr
½ quint 9 ß 10 hlr
1 d 4 ß 11 hlr

Holt zu 19 karat
1 marck 66 fl 10 ß

After a few blank pages a new subject begins on fol 48r: warfare. This chapter contains a large number of illustrations, mainly of cannons, and a text on the defense of a castle which includes some pyrotechnical recipes. It covers three gatherings, thus corresponding in size to the chapter on the extraction of metal and its use in coinage.

There had been an explosive spread of illuminated manuscripts on the technology of war since the beginning of the century. Just after 1400, Konrad Kyeser of Eichstätt assembled a plainly drawn book of machines based on models from antiquity. He entitled it *Bellifortis* ('he who is strong in battle') as it deals mainly with war machines, and many copies of it (fig. 53) have survived. It was a great success up to the middle of the century. The most famous copy is a luxury manuscript now in Göttingen, the only one with a portrait of the author.[38] The *Bellifortis* subdivided war technology in chapters according to different criteria, whether it was for attack, defense, or concerned with fire or water. Kyeser prefaced his book with the mounted planetary gods who determine the course of events in the world (fig. 23). This led over to a portrait of Alexander the Great as the ideal ruler, well versed in technology. Not exactly modest, Kyeser designated his *Bellifortis* as the Bible for secular rulers, on a par with the Scriptures for the clergy.

After 1500 the *Bellifortis* lost its attraction. The prime of works on technology that combined war machines with clothes irons and pocket knives in one volume was also over. Furthermore, firearms had now become the modern method of waging war and the new types of cannons would have gone beyond the scope of a book. Two codices of immense proportions demonstrate the futility of such endeavor. The Frankish knight Ludwig von Eyb the Younger (1450-1521), a very successful court official, wrote several books dealing mainly with chivalry and the aristocracy. His book on war is largely a review of older sources, including the Bellifortis, and comprises an enormous total of about 600 pages.[39] In the same category is the so-called *Ingenieurkunst- und Wunderbuch (Book of Engineering Arts and Wonders)* now in Weimar. Though on parchment and with additional illustrations of professions and magic, it numbers an almost identical 329 folios.[40]

The source for the drawings in the Housebook was definitely not the *Bellifortis* itself, but both books depend on the same tradition of works on war technology which is much older than the *Bellifortis*. Kyeser's book is merely the first tangible evidence of it because hardly anything from an earlier period has survived. Moreover, research on the links to antiquity is scanty. Typical motifs in such books are the movable walls serving to protect the gunners loading the cannon.

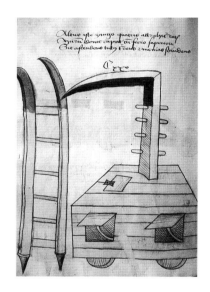

53   *Battle wagon with a ladder and a wall claw in a 'Bellifortis' manuscript, c. 1440, Munich, Bayerische Staatsbibliothek*

Fol. 52v shows a cart with protective walls. It is pushed by six horses and has a gun mounted on the front (fig. 54).

The 'covert art' of scaling enemy walls was a favorite subject in these specialized illuminated manuscripts. Dozens of variants of ladders – dismantlable ladders, rope ladders, ladders with one or more cross or side beams – are usually lined up unimaginatively, rarely demonstrating how they are used. In contrast, the Housebook integrates the scaling gear into a charming composition. A maiden waits in the window of a tall tower, from which a youth is descending in the manner of a mountaineer, while two others climb up, one with the aid of a jousting lance, the other with climbing irons and a taut weighted rope (fig. 55). The scene is flanked by the usual representations of a rope ladder and a tripartite wooden ladder.

While none of the Housebook's drawings can be traced directly to the *Bellifortis*, the war engines are in many cases the same, such as the extendable ladders and the arquebusier's shield. The two works are related through their common subject. Their drawings, though both soberly drawn, are stylistically quite different.

More important than the *Bellifortis* as a parallel to the Housebook is a selection of realistic drawings of war technology of Nuremberg provenance, appearing in the middle of the 15th century (fig. 57).[41] It fed into many compilations and this corpus of manuscripts was misleadingly named after the author of one of them: Johannes Form–

54 *Battle wagon with chassis, fol. 52v*

Overleaf:
55 *A lathe, support, plumb line level, and scaling equipment, fols 53v-53v1*

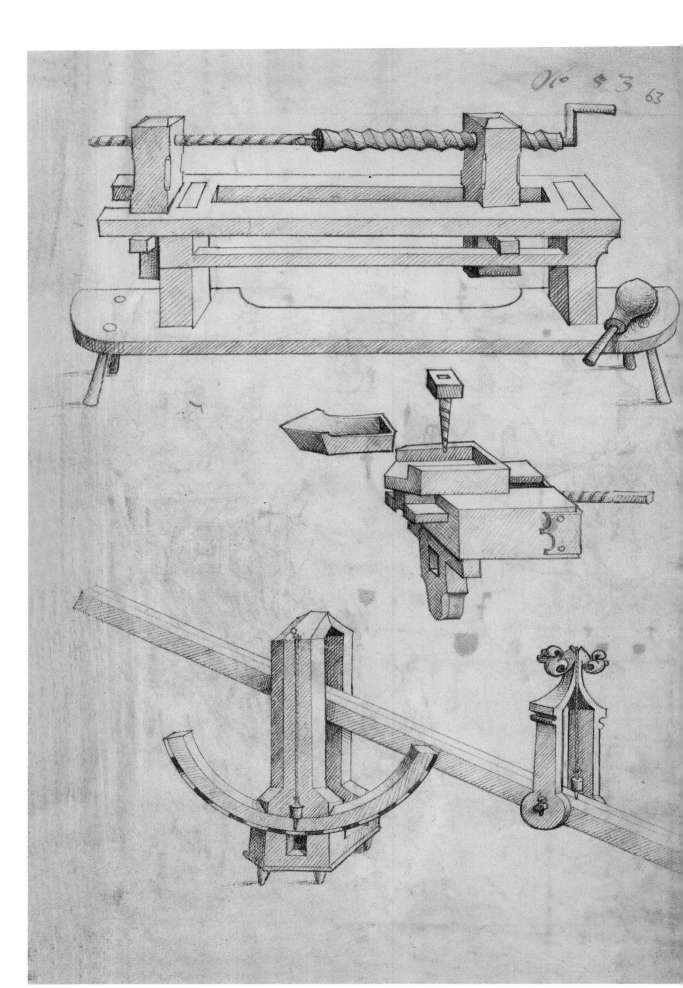

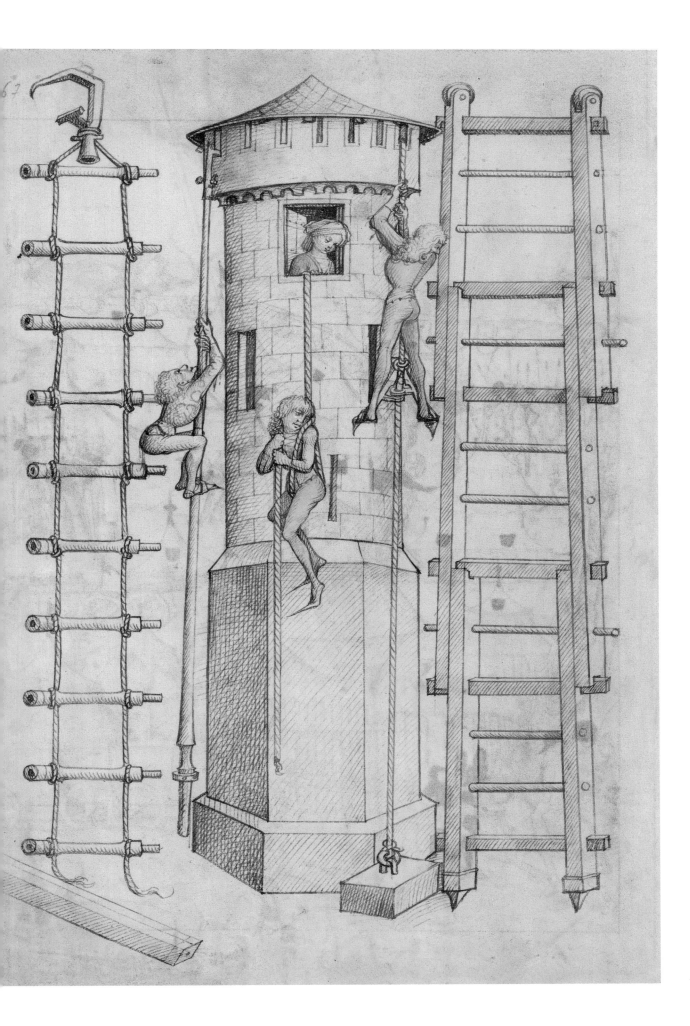

schneider, an artillery master who was granted citizen's rights in Nuremberg in 1440. Many of the machines in the Housebook are related to the Formschneider group. A manuscript in the Bavarian State Library in Munich (Cgm 356) is especially close, with six illustrations almost identical to those in the Housebook.

Complicated machinery, such as the mechanics of mills (figs 56 and 63), is depicted with great artistic skill. All parts are easily identifiable while the perspective remains credible at the same time. The handmills themselves were often built such that they could be dismantled and transported on supply carts. A plumb level, which doubles as a plumb quadrant for setting up cannons, appears on several pages (fig. 55).

The technical section originally formed a larger part of the manuscript as a whole than it does today. Six leaves have been shown to

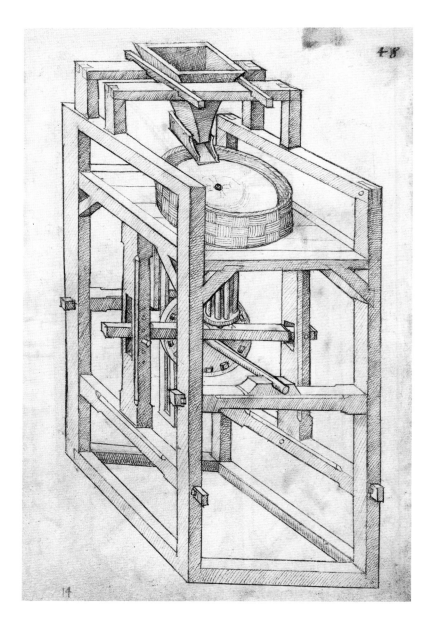

56   *Handmill with pin gear, fol. 48r*

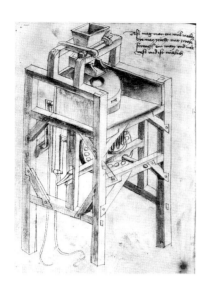

57 *Handmill in the 'Henntz Codex,'*
c. 1450 (?), Weimar, Herzogin Anna
Amalia Bibliothek, Q 342

be missing before the drawing of a hurling apparatus in which tension is attained by winding down a curved beam (fol. 56r). Perhaps they illustrated other hurling machines.

A treatise for an artillery master is still extant. It begins on fol. 57r with the words: "Item this is right and proper for a munitions master. He should keep God in mind when he is dealing with gun and powder." However, the tone and contents soon change, and address the reader directly: "Item if you are in charge of organizing a castle." This is followed by practical advice that should be observed by a castellan in running a castle. It begins with criteria for the selection of subordinates and with tips on how to deal with them: "Know and make sure that you serve enough food and drink so that your subordinates have no cause for complaint or grudge against you."

The man in charge must among other things look after defense equipment, including the construction of embrasures, the supply of stones to hurl down from the battlements, and the organization of an efficient watch. Instructions to follow in an enemy attack are next. If the situation requires that all cannons be fired at once, rocks and pots filled with caltrops should be hurled down on the enemy as well in order to prevent him from approaching during pauses in firing while the guns are being loaded. If ladders have already been set up – we have seen them depicted earlier in the Housebook – fire and boiling water are of help. Should the enemy be digging a tunnel, you listen where it is, dig towards him, light a fire and smoke him out.

That the technical drawings in the Housebook are the work of an outstanding artist becomes clear by comparison with the other manuscripts of the period on war technology. Not a single one of them succeeds in displaying the mechanisms as consistently clearly while disposing them in such pleasing and well proportioned fashion on the page. Many fail because they are unable to adapt their models. Thus blank spaces remain where the text was not taken over as well, or else the form was copied without any understanding of the function. Providing a typical though later example, a Frankfurt manuscript depicts a bag-shaped battle chariot known as a 'monk's hood' in the *Bellifortis* in two completely different ways.[42]

The Army on the March and the Encampment

A large panorama of an army on the march extends across two bifoliate foldout pages, three full-length rows of carts travelling across their width (figs 58, 60a and b). The middle row transports provisions and military equipment. It is guarded by the two outer rows of battle wagons equipped with defensive walls with openings through which guns can fire. One of these can be seen on the verso, clearly showing that there is a swivel-mounted light cannon inside (fig. 54). The fluttering standard at the end of the train bears the letters "A E I O [U]," the device of the Habsburgs, particularly of Frederick III.

59   *Gun cart and gun wagon with walls*
and swivel-mounted gun, fol. 52v1

In this illustration the Housebook keeps within the normal bounds, while a march in a book by the Palatine munitions master Philipp Mönch, for instance, is so long it has to wind back and forth across the page to fit.

Like the scaling apparatus, the barricade of wagons around the encampment is part of the conventional repertoire of war manuals (fig. 61). As in the charming scene of the men scaling the tower, motifs in the foreground such as the horse's cadaver or the card players are additional details which, here too, attract attention. Much effort has gone into making this barricade formation more interesting. It is less of an historical view than a scheme for the best way to array an army camp, protected by concentric rings of guncarts. This provided the opportunity for a lively portrayal, ranging from beggars at the camp gate to military leaders gathering around the imperial banner (fig. 62).

Heraldry lends the camp a further appearance of reality. The arms indicate who is present: the emperor and lords whose rank is expressed by the size of their shields and banners. The most magnificent is the imperial banner; next in rank is the one with the device of the lion and wheel of his chancellor Adolf II of Nassau, who was also Bishop of Mainz (fig. 5). Very small in the background is a tent with shields displaying the arms of the tree trunk. Beyond the inner circle appears the star device of the Erbach family from the Odenwald, who were cup bearers of the Count Palatine. Another shield, with two rows of antlers, cannot be identified (three would have been Württemberg).

Overleaves:
60a and b   The army on the march,
fols 51v-52r1

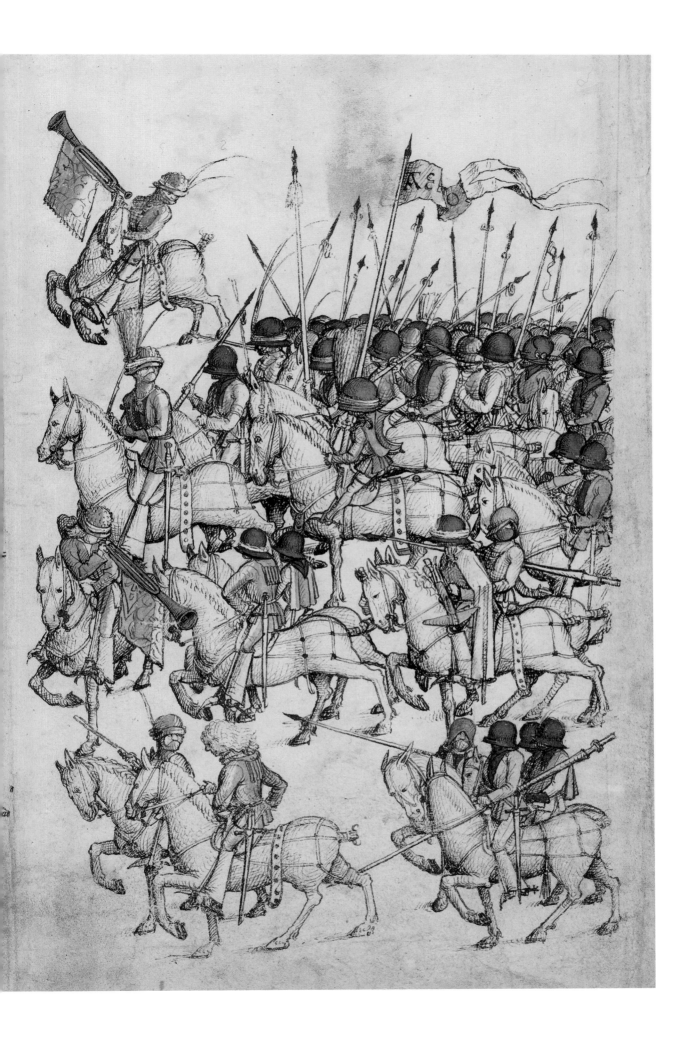

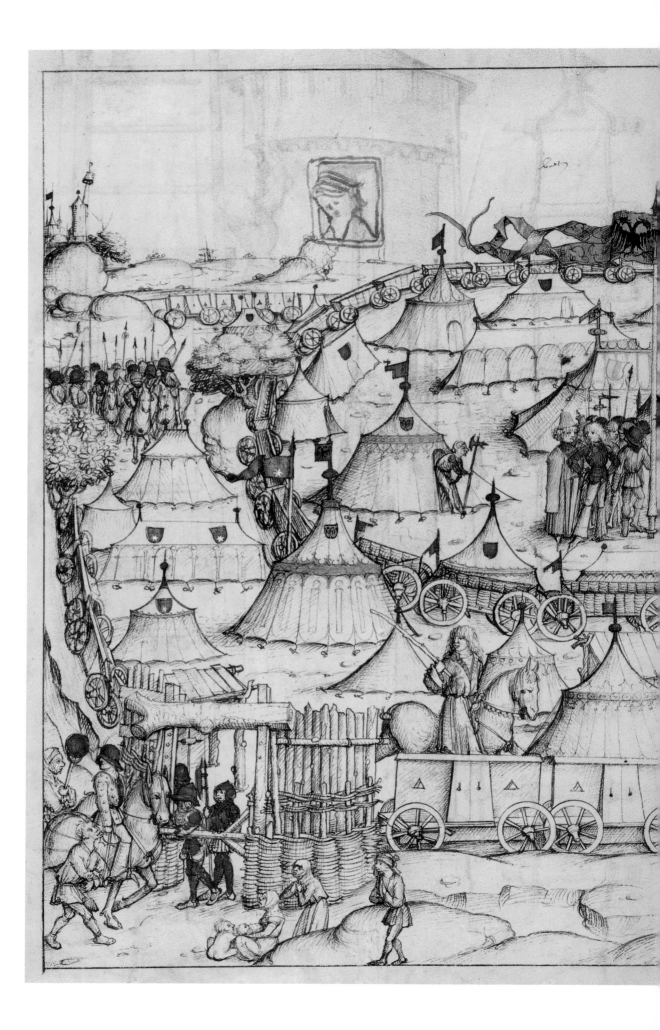

A chevron device shown in two different color schemes could be attributed to Hanau as well as to other families. The rest of the arms are unidentifiable.

At best these arms might indicate in what circles the patron moved, which geographical area and what families were important in his life. The military campaign represented is, however, not an historical one. Research tried to identify it with the encampment near Neuss in 1475. Rather, it is an invention developed through the narrative embellishment of a military scheme.

62   *The military leaders, detail from the army camp, fols 53r-53r1*

Preceding facing pages:
61   *The army encampment, fols 53r-53r1*

Summary

All the manuscripts with technical drawings comparable to the House-book originated in the period ranging from the middle of the 15th century to just after 1500. Direct copies of the *Bellifortis* drawings are not to be found in the Housebook. While the large panoramas are conventional, both the march and the camp take advantage of opportunities to make the demonstration of tactical know-how as vivid as possible. Unusual for the genre is only the solution of using a foldout. It saves having to divide the train into several rows for it to fit onto the page.

No space has been left on the folios for any explanatory text. The drawings speak for themselves like modern engineering drawings and presuppose a considerable technical background on the part of the beholder. A layperson may just manage to understand the war technology that is not too complicated, but the metallurgy section makes much greater demands. It requires specialized knowledge. The illustrations alone, for example, scarcely explain the method of extracting silver by smelting with the furrow drain process.

The Housebook is no isolated monolith in the landscape of manuscripts. Almost all of its contents belong to a familar repertoire: war and mining technology, the planets and the liberal arts, even in combination with prescriptions and recipes. What is without parallel in a miscellany, besides its outstanding artistic quality, is that the items of interest have not been arbitrarily assembled, as was often the case in contemporary manuscripts, but that an attempt to create order pervades the book. In the case of Kyeser's book, its inner cohesion, though lost in later copies, is still discernible. Other genres such as the world chronicles, for instance the *Schedelsche Weltchronik* which appeared in Nuremberg in 1493, continued to be embedded in the ordering principle of a God-fearing system, beginning with the Creation and ending in the Last Judgement.

Even in the present state of the Housebook its underlying structure is still recognizable. Antithesis is a favorite means to this end, as it was in the medieval battle of the virtues and vices or in the way typology linked Old and New Testament antetypes and types. The Housebook does not make any obvious didactic or moralizing statements in the scenes of knightly life. The bathhouse or the martial joust imply no negative message. What does attract notice is the opposition between an ideal imaginary world and practical reality. The garden of love, the playful tournament and the noble stag hunt appear like a dream.

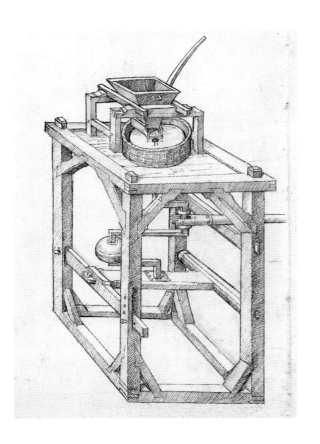

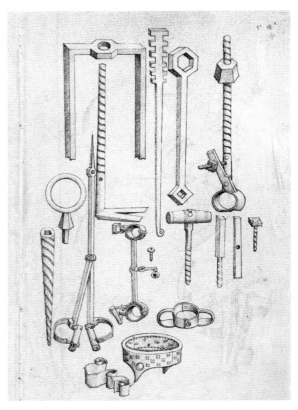

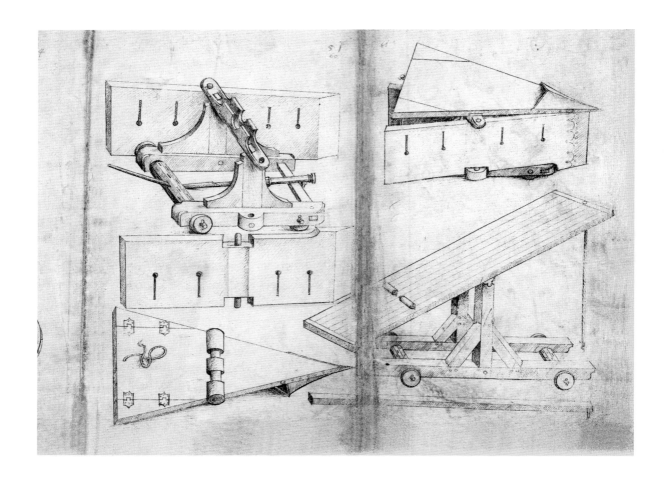

Reality consists of the life around a castle, of the joust – not as mere entertainment but as training for war – and finally, of wanton wooing, here given an ironic touch: it is not the men who are courting according to established rules, but the 'fair sex' is catching the men with cunning and dirty tricks. The chivalrous ideal is contrasted with a persiflage of reality.

It is very tempting to match the program of the Housebook as a whole with one of the well known medieval schemes. The system of the 'artes' which organizes knowledge suggests itself. Konrad Kyeser's codex was itself based on it. If the art of memory and astrology form a kind of intellectual superstructure for the rudimentary remains of the first part of the book, the refining of metals and their use as money could be intended as prerequisites for waging war in the second part. The latter, however, lacks such elaborate references. It contains the chapters on smelting, mainly chemical procedures, followed by mintage and finally the technology of war. The distinct division of the Housebook in the middle and the scene of the fight on the central page showing mining also support this hypothesis. Another possible program is the succession from the liberal art of memory down to the mechanical art of war.

Even though we cannot unclose any scheme to our total satisfaction, and the key of the 'artes' turns but stiffly in the lock, the exceptional presence of structure in the Housebook at least remains evident – both as a whole and within the chapters, especially in the series of courtly scenes with their highly intricate and witty play of mutual correspondences.

THE PATRON

Beyond the mere systematics of technology, the Housebook extends into several other domains. On the one hand it presents a very traditional view in the courtly ideals, and on the other great modernity in the technology. The dichotomy this entails becomes apparent only with hindsight. The fashionable gentleman in pointed shoes was on his way out, to be replaced by a completely new kind of Renaissance portrait. Confronting the old pictorial world with innovation makes seem it more antiquated than it was. In fact, chivalrous ideals and the imaginary world of love allegories continued to be associated with an image of belonging to the upper classes long into the next century. This image was increasingly appropriated by the bourgeoisie, now upwardly mobile after the major social upheavals of the age. What do we

Top left:
63   *Handmill with counterbalance weight, fol. 48v*

Top right:
64   *Battering and locking instruments, fol. 50r*

Below:
65   *Shielding devices for guns, fols 51r-51r1*

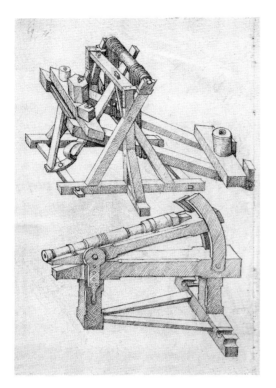

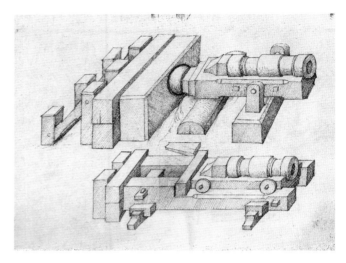

66   *Catapult and gun, fol. 54v*

67   *Block mounts, fol. 54α*

68   *Three turret guns, fol. 50v*

69   *Guns on carriages, fol. 55r*

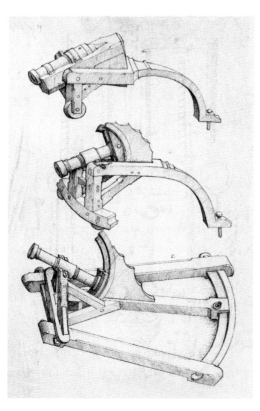

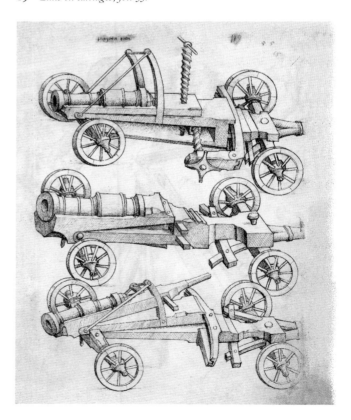

know about the patron of the Housebook besides his coat of arms? He was presumably a member of the aspiring bourgeoisie. Very probably he had been granted the Order of the Jug and was close to the court. This is suggested by the encampment scene and by the fact that it was the Housebook Master he gave the commission to. This artist was active in court circles, as suggested among other things by a sheet with two drawings attributed to him showing King Maximilian in Bruges.[43]

In a certain sense the Housebook belongs in a completely different category from the *Bellifortis*. Apart from its personal associations with the author, the latter has little relation to its purchasers. It reflects a development that was later continued by the blockbooks and works printed with movable type: an increasing anonymity. The days of having a prayer book put together with your favorite devotions or having various pieces of prose bound in a volume were nearing their end. It was not long before the only individual feature left to give a book was the binding, matching the others in your private library. By contrast, the Housebook's purpose and aim are the representation and glorification of the patron. Similarly, the book made for Marx Walther of Augsburg mentioned above contains, besides tournaments, a genealogy, a family tree and a list of endowments. It proclaims the fame of the owner and his lineage. What is encompassed by the collected works of Ludwig Eyb the Younger, with a book on heraldry, a tournament book and the technical book introduced above, is a world of chivalry and nobility that is combined in the Housebook in one volume. A didactic tone prevails in Ludwig Eyb's work: "as a lesson, an example to all young knights," he writes in his *Wilwolt*. In his war manual he addresses those "who are inclined toward noble knightly good deeds." The Housebook, begun some twenty years earlier, likewise adapts forms suited to establish the social status of its owner. These works contain much that reflects notions of the ideal held by the hereditary upper classes. This included courtly love. Technical matters were not dealt with for their own sake but subordinated to a higher purpose.

Whether the original owner of the Housebook will ever be found is doubtful. To say the least, he was neither a munitions master nor a miner, but someone with a sense for the complex relationships which lend books their own cosmos in the good old traditional manner. He was more at home in the world of the doctor and humanist Hartmann Schedel than that of the artillery master Philipp Mönch. He was an educated man who took the time and trouble to assemble material for a book. We cannot tell whether his coat of arms exists only in the Housebook and forms part of its design. Yet one thing seems certain: he was a Knight of the Order of the Jug and as such he conducts us through his book.

As mentioned at the outset, the question of the identity and number of the artists who contributed to the Housebook is highly controversial. Assuming that the Housebook is not an arbitrary conglomerate, as we have demonstrated here, the hypothesis that various artists participated is difficult to sustain.

I would basically argue for a single artist instead of several. The children of the planets and the scenes of knightly life are by the same artist, despite the unmistakable differences in composition of the planet drawings. They enumerate the various children of the planets from Saturn to Luna side by side, thus following the tradition that required each to be individually mentioned and clearly visible (figs 15–21). A blockbook need not have been the source; there could have been another, unknown to us. These pictures only had to list what was in the text, presenting various human characteristics without evaluating them. Nevertheless, the development of spatial relationships can be traced as they proceed. In an attempt to create the impression of depth, the gloomy children of Saturn are layered in rows above instead

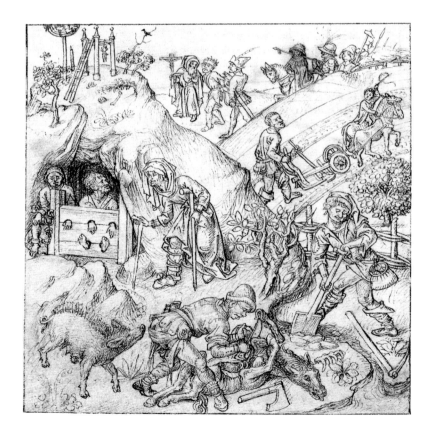

70   *The children of Saturn, detail, fol. 11r*

of behind each other (fig. 70). Spatial relationships are lacking; the figures move on different planes and will never meet on the page. The restless and water-loving childen of Luna, on the other hand, inhabit the landscape together. A pond joins up the scenes and directs the view towards a lake in the distance (fig. 71). The price for this convincing perspective is paid by the fishermen. Pushed off into the distance, they cast their nets, playing only a minor role in the circle of Luna's children compared to the itinerant conjurer and his audience.

In the depictions of knightly life the differing point of departure is decisive. There was no model for them, no scheme on which the artist could base himself. The drawings betray his difficulties in his efforts to invent totally new compositions. As he played his sophisticated game of cross references with repetitions and mirroring, the harmony within the individual scenes was sacrificed. Developing a new scheme is much more difficult than falling back on traditional material. In the planet drawings the artist could expand and revise illustrations from the blockbook tradition, which he was not able to do in the others. Correspondingly, scenes such as the interior of the smelting plant with the furnaces in the mining chapter have a few shortcomings: apart from problems with perspective, the elegant visiting couple is rather stiff.

71   *The children of Luna, detail, fol. 17r*

72 *Activity during the march of the army, detail, fol. 52r*

Furthermore, the friendly, playful sense of humor we encounter everywhere speaks against the collaboration of several artists. The vain maiden as the sign of the zodiac in Mercury (fig. 73) or the fish in Jupiter – a pike devouring a stickleback – are witty innovations. It is scarely conceivable that two different artists could correspond this fully in personality and skill, unless they were brothers.

The same refreshing touch is apparent in the encampment and the march. The technical matters they represent determine the composition. But while the structure of military order has priority, the scenes are livened up – here by card players, there a poodle – witnessing to the same delight in comic narrative. The artist has a weakness for dogs; several different breeds are represented in the Housebook. Horses, on the other hand, are routine for him. He draws them with an experienced hand, almost stereotype. The other drawings of war technology are presumably also by him, for their difference in artistic quality from most other books on the subject is remarkable.

The debate about the artist, not to mention the book's relationship to other works by the Housebook Master, is far from over. But one thing is certain: the decision of the patron to choose this outstanding artist when he commissioned his book was a great stroke of luck.

73   *Virgo, the sign of the zodiac, detail of fol. 16r*

Notes

1   Retberg 1865, p. 4.

2   Heidelberg, Universitätsbibliothek MS Cod. pal. germ. 87.

3   Hess 1994, p. 15 ff.

4   For detailed information on Maximilian Willibald's collection see Bernd M. Mayer, *Von Schongauer zu Rembrandt. Meisterwerke der Druckgraphik aus der Sammlung der Fürsten zu Waldburg-Wolfegg*, Ravensburg 1996, p. 9 ff. On the man himself, see Peter Eitel, *ibid.*, p. 20 ff.

5   Essenwein 1887.

6   Bossert/Storck 1912.

7   Waldburg 1957.

8   *Vom Leben im späten Mittelalter* 1985.

9   Waldburg 1997.

10   *Meister E. S.* 1987, no. 72, p. 64.

11   See also *Spielkarten – ihre Kunst und Geschichte in Mitteleuropa*, exhib. cat. Graphische Sammlung Albertina, Vienna 1974, no. 21, p. 64 ff.

12   Gustav Hergsell, *Thalhofers Fechtbuch aus dem Jahre 1467. Gerichtliche und andere Zweikämpfe darstellend*, Prague 1887.

13   *Meister E. S.* 1987, no. 98, p. 83 f.

14   *Ibid.*, 'Nativity,' no. 10, p. 28.

15   Rainer Leng, 'The Prelude to the Housebook: Mnemosyne,' in: Waldburg 1997, p. 113 ff.

16   For an outline of medieval lay astrology and complete bibliography, see Gundolf Keil (ed.), *Vom Einfluss der Gestirne auf die Gesundheit und den Charakter der Menschen. Kommentar zur Faksimile-Ausgabe des Manuskripts C 54 Zentralbibliothek Zürich (Nürnberger Kodex Schürstab)*, Lucerne 1981-82. For a clear description of the history of astronomy see Jürgen Teichmann, *Wandel des Weltbildes. Astronomie, Physik und Messtechnik in der Kulturgeschichte*, 3rd edn, Stuttgart and Leipzig 1996.

17   Alexander Neckam, *De naturis rerum*, ed. T. Wright, London 1863, p. 39 ff.

18   English translations of the verses are by Marianne Hansen, reproduced, with minor changes, with kind permission.

19   *Blockbücher des Mittelalters. Bilderfolgen als Lektüre*, exhib. cat., Gutenberg Museum, Mainz 1991, p. 200 ff.

20   Aschaffenburg, Hofbibliothek MS 12, dating from between 1464 (Adolf II's entry on office) and 1475. For drawing style and manuscript illumination in the Housebook, see also Eberhard König, 'The Housebook Master,' in: Waldburg 1997, p. 163 ff.

21   Hans Peter Duerr, *Nacktheit und Scham* (Der Mythos vom Zivilisationsprozess, vol. 1), Frankfurt am Main 1988. Duerr discusses Norbert Elias, *Über den Prozess der Zivilisation. Soziogenetische und psychogenetishe Untersuchungen*, vol. 1: *Wandlungen des Verhaltens in den weltlichen Oberschichten des Abendlandes*, 2nd edn, Berne and Munich 1969, p. 285. Elias uses several scenes in the Housebook as evidence for the uninhibited way that 'natural' bodily functions were dealt with in the Middle Ages. In opposition, Duerr shows that the children of Venus in the Housebook are far removed from real life but stick close to the text on the planet, and he demonstrates how proper bathing in the Middle Ages could be; pp. 34-58.

22   For tournament procedures, see Ortwin Gamber, 'Ritterspiele und Turnierrüstung im Spätmittelalter,' *Das ritterliche Turnier im Mittelalter*, ed. Josef Fleckenstein, Göttingen 1986, pp. 513-531.

23   Heinrich Mang or Lang. For bibliography on this subject, see Jane Campbell Hutchison, *The Master of the Housebook*, New York 1972, p. 82.

24   Elmar Mittler and Wilfried Werner (eds), *Codex Manesse. Die Welt des Codex Manesse. Ein Blick ins Mittelalter* (Heidelberger Bibliotheksschriften, 30), exhib. cat. Universitätsbibliothek Heidelberg 1988.

25 Fragment, 47 X 45 cm, private collection. Illustrated in Hess 1994, fig. 46, p. 53; see also p. 149 ff.

26 Quoted from Richard Froning, *Frankfurter Chroniken und annalistische Aufzeichnungen des Mittelalters. Quellen zur Frankfurter Geschichte*, vol. 1, Frankfurt am Main 1884, p. 221.

27 Munich, Bayerische Staatsbibliothek MS Cgm 1930; Arnd Reitemeier, 'Marx Walther: Turnierbuch,' *"Kurzweil viel ohn' Mass und Ziel." Alltag und Festtag auf den Augsburger Monatsbildern der Renaissance*, ed. Pia Maria Grüber, exhib. cat. Augsburg, Munich 1994, p. 186 f.

28 Stained glass roundel with imperial arms, New York, The Metropolitan Museum of Art, The Cloisters Collection, inv. no. 11.120.2; *Gothic and Renaissance Art in Nuremberg 1300-1550*, exhib. cat. Nuremberg and New York, Munich 1986, no. 66, p. 206 f. Another roundel of Nuremberg provenance, Baltimore, Walters Art Gallery, inv. no. 46.76; *ibid.*, no. 264, p. 454 f.

29 Ingeborg Glier, *Artes amandi. Untersuchung zu Geschichte, Überlieferung und Typologie der deutschen Minnereden* (Münchner Texte und Untersuchungen zur deutschen Literatur des Mittelalters, ed. Kommission für deutsche Literatur des Mittelalters der Bayerischen Akademie der Wissenschaften, vol. 34), Munich 1971, pp. 253-256.

30 *Das Stuttgarter Kartenspiel*, partial facsimile edn, ed. and intro. Heribert Meurer, Stuttgart 1979.

31 Quoted from Glier (as in n. 29), p. 157.

32 Anna Coreth, 'Der "Orden von der Stola und den Kanndeln und den Greifen" (Aragonesischer Kannenorden),' *Mitteilungen des Österreichischen Staatsarchivs*, V, Vienna 1952, pp. 34-51.

33 Haus-, Hof- und Staatsarchiv (Royal, Imperial and State Archive) Vienna, copy B1, MS Böhm 107, blue 43, fols 69-70v; *ibid.*, p. 56 ff.

34 *Ibid.*, p. 40 f.

35 *Ibid.*, p. 42 f.

36 *Ibid.*, p. 34.

37 For an introduction to the complicated procedures of smelting see Wilfried Liessmann, *Historischer Bergbau im Harz. Ein Kurzführer* (Schriften des Mineralogischen Museums der Universität Hamburg, vol. 1), Cologne 1990, p. 95 ff. See also Karl-Heinz Ludwig, 'Historical and Metallurgical Observations on the Medieval Housebook,' in: Waldburg 1997, p. 127 ff.

38 Götz Quarg (ed.), *Conrad Kyeser aus Eichstätt. Bellifortis. Faksimiledruck der Pergament Handschrift Cod. ms. philos. 63 der Universitätsbibliothek Göttingen*, facsimile edn., Düsseldorf 1967, 2 vols; see major corrections in a review by Hermann Heimpel, in: *Göttingsche Gelehrte Anzeigen*, 223, no. 1,2, 1971, p. 115.

39 Universitätsbibliothek Erlangen-Nürnberg MS B 26; Hans-Otto Keunecke, 'Ludwig von Eyb d. J. zum Hardenstein, Kriegsbuch,' *Cimelia Erlangensia – Aus den Schätzen der Universitätsbibliothek* (Schriften der Universitätsbibliothek Erlangen-Nürnberg, 24), exhib. cat. Erlangen 1993, no. 26, pp. 63-65.

40 Weimar, Herzogin Anna Amalia Bibliothek Fol. 328; Konrad Kratzsch, *Kostbarkeiten der Herzogin Anna Amalia Bibliothek – Weimar*, Leipzig 1993, pp. 206-211.

41 The earliest (mid-15th-century) example is the so-called 'Henntz Codex' in Weimar, Herzogin Anna Amalia Bibiliothek Q 342. War technology in the Housebook is discussed by Rainer Leng, '"Burning, killing, treachery everywhere / Stabbing, slaying in fiercest war". War in the Medieval Housebook,' in: Waldburg 1997, p. 145 ff.

42 Frankfurt am Main, Universitätsbibliothek MS germ. qu. 14 (fols 108r and 123r).

43 *Dürer, Holbein, Grünewald. Meisterzeichnungen der deutschen Renaissance aus Berlin und Basel*, exhib. cat. Berlin and Basle 1997, no. 3.2, pp. 43-45.

Selected Bibliography

Bossert / Storck 1912
Helmuth T. Bossert and Willy F. Storck, *Das Mittelalterliche Hausbuch nach dem Originale im Besitz des Fürsten von Waldburg-Wolfegg-Waldsee*, Leipzig 1912

Essenwein 1887
August von Essenwein (ed.), *Mittelalterliches Hausbuch – Bilderhandschrift des 15. Jahrhunderts mit vollständigem Text und facsimilierten Abbildungen*, Leipzig 1866; 2nd edn, Frankfurt am Main 1887

Hess 1994
Daniel Hess, *Meister um das 'mittelalterliche Hausbuch.' Studien zur Hausbuchmeisterfrage*, Mainz 1994

Lehrs
Max Lehrs, *Geschichte und Kritischer Katalog des Deutschen, Niederländischen und Französischen Kupferstichs im XV. Jahrhundert*, Vienna 1908-1934, 9 vols

Meister E. S. 1987
Holm Bevers, *Meister E. S. Ein oberrheinischer Kupferstecher der Spätgotik*, exhib. cat. Munich and Berlin 1987

Rathgen 1928
Bernhard Rathgen, *Das Geschütz im Mittelalter*, new edn, ed. and intro. Volker Schmidtchen, Berlin 1928, repr. Düsseldorf 1987

Retberg 1865
Ralf von Retberg, *Kulturgeschichtliche Briefe. Über ein mittelalterliches Hausbuch des 15. Jahrhunderts aus der fürstlich Waldburg-Wolfeggischen Sammlung*, Leipzig 1865

Schefold 1929
Max Schefold, 'Das mittelalterliche Hausbuch als Dokument für die Geschichte der Technik,' *Beiträge zur Geschichte der Technik und Industrie – Jahrbuch des Vereins deutscher Ingenieure*, 19, 1929, pp. 127-132

Sterzel 1912-14
H. Sterzel, 'Das Wolfegger Hausbuch und seine Bedeutung für die Waffenkunde,' *Zeitschrift für Waffenkunde*, 6, 1912-14, pp. 234 ff, 280 ff, 314 ff

Vom Leben im späten Mittelalter 1985
Jan Piet Filedt Kok (ed.), *Vom Leben im späten Mittelalter. Der Hausbuchmeister oder der Meister des Amsterdamer Kabinetts*, exhib. cat. Rijksmuseum Amsterdam and Städelsches Kunstinstitut, Frankfurt am Main 1985. English edition: *Livelier than Life. The Master of the Amsterdam Cabinet or the Housebook Master, ca. 1470-1500*, Amsterdam 1985

Waldburg 1957
Johannes Graf zu Waldburg Wolfegg, *Das mittelalterliche Hausbuch – Betrachtungen vor einer Bilderhandschrift* (Bibliothek des Germanischen National-Museums Nürnberg zur deutschen Kunst- und Kulturgeschichte, vol. 8), Munich 1957

Waldburg 1997
Christoph Graf zu Waldburg Wolfegg (ed.), *Das Mittelalterliche Hausbuch*, facsimile edn. and commentary, Munich 1997, 2 vols

Photo credits

Pictorial material used for reproduction is the property of the owners of the works or from the author's archives.
Stiftung Weimarer Klassik: fig. 57.
Fürstlich zu Waldburg-Wolfegg'sche Kunstsammlungen, Wolfegg:
fig. 3 and the Medieval Housebook.
The folios of the Medieval Housebook were photographed
by Engelbert Seehuber, Munich.